DEVELOPING COUNTRIES AS EXPORTERS OF TECHNOLOGY

By the same author

FOREIGN PRIVATE MANUFACTURING INVESTMENT AND MULTINATIONAL CORPORATIONS: An Annotated Bibliography
FOREIGN INVESTMENT, TRANSNATIONALS AND DEVELOPING COUNTRIES (*with Paul Streeten*)
THE GROWTH OF THE PHARMACEUTICAL INDUSTRY IN DEVELOPING COUNTRIES
THE MULTINATIONAL CORPORATION: Nine Essays
DEVELOPING COUNTRIES IN THE INTERNATIONAL ECONOMY

DEVELOPING COUNTRIES AS EXPORTERS OF TECHNOLOGY

A First Look at the Indian Experience

Sanjaya Lall
Oxford University Institute of Economics and Statistics

To
LALIT
a dear friend

First published 1982 by
THE MACMILLAN PRESS LTD
London and Basingstoke
Companies and representatives throughout the world

ISBN 0 333 28844 0

Printed in Hong Kong

Contents

Preface

This book presents a preliminary evaluation of the emergence of some of the more industrialized developing countries as internationally competitive sellers of technology. It develops the analysis of an earlier paper (Lall, 1979), but it devotes particular attention to India which, despite its overwhelming poverty and a relatively poor record of economic growth, seems to lead the Third World in this activity.

The phenomenon of technology exports by developing countries is itself worthy of note. The fact that they have established a comparative advantage in the sale of the most skill and knowledge-intensive of all products – pure know-how itself – raises important issues for their dynamic role in international trade. It also calls for a closer examination of the processes of technological assimilation and development which underlie their entry into foreign markets: which sectors and enterprises have shown the greatest technical progress, whether or not technologies have been adapted to local needs, the role which imports of technology have played, and the significance of government policy in promoting or inhibiting technological development. If technical growth is indeed the motor of economic growth, these questions are central to large areas of analysis and policy formulation.

The fact that different developing countries are revealing different patterns of comparative advantage in exporting technology raises a further set of interesting issues. The fact, in particular, that the poorest and least successful (at least in conventional terms of growth of per capita incomes) of the so-called 'newly industrializing countries', and the one which still pursues the most inward-looking strategy, is the most versatile and advanced technology exporter leads us to examine again the current orthodoxy on the almost unmitigated harms of protected industrialization, and on the belief that high levels of skill and technology are always associated with high levels of income.

This monograph raises more questions than it answers. It is a 'first look' in both of two senses: first, the data on which it is based are drawn from a variety of scattered sources, and they are incomplete and patchy. The very novelty of the phenomenon forces one to rely on newspaper reports, impressions and the occasional official survey for the primary information.

Second, the nature of the economic processes involved is not well understood. The hypotheses which can be advanced at this stage must perforce be tentative, and should be treated as such. Until detailed fieldwork is undertaken to map out the true extent of technology exports and, more importantly, to chart the course of technological change in the enterprises involved, this is as much as can be achieved from the outside. Provided its limitations are borne in mind, there is a useful part to be played by such studies in stimulating debate and research.

The original draft on which the present work is based was prepared for the Commonwealth Secretariat, and I am deeply indebted to the Secretariat for allowing me to adapt it for the purposes of publication. Since that draft was written, it has become possible for me to undertake further research into technology exports by a number of developing countries on behalf of the World Bank and the Inter-American Development Bank. The results of this research, if it materializes, will not be published for some years from now. I have decided, therefore, to publish this preliminary version now, being aware that some of its conclusions may be modified in the light of future findings.

My thanks are due to a large number of people who commented on earlier drafts and discussed the subject with me at length. I would like to acknowledge my gratitude, in particular, to Charles Cooper, Mariluz Cortez, Bimal Jalan, Prem Jha, Keith Pavitt, Ashwani Saith, Raymond Vernon, Larry Westphal and Martin Wolf. Jorge Katz, who has conducted pathbreaking research into the subject of technological development in Argentina, provided me access to his work and thus furnished much of the intellectual underpinnings of my analysis. Richard Thomas of the Foreign Office gave me his own collection of data on Indian technology exports. J. L. Bajaj and Usha Dar of the Indian Investment Centre sent me information on Indian direct investments abroad. A number of others, at various seminars and meetings, gave me the benefit of their insights and knowledge. To all these, my thanks. Needless to say, they bear no responsibility for the interpretation that I have placed on the facts and arguments they provided.

My wife suffered patiently through innumerable and repetitious discussions of technology exports. For this, and for very much more, I am eternally indebted to her. Gillian Coates, Sybil Owen, Caroline Wise and Caroline Baldwin typed the drafts with great speed and efficiency – I thank them for their help.

Oxford, 1979 S.L.

Part A

Significance and Definitions

1 Introduction: Significance of Developing Country Exports of Technology

A number of developing countries with experience of industrialization are emerging on the international scene as exporters of manufacturing, construction, management, financial and other forms of technology. Those that possess local capital-goods manufacturing sectors are able to export the technical know-how and equipment required to set up a range of manufacturing industries abroad, the leading examples being India, Argentina, Brazil, México, South Korea and Taiwan.[1] Those that have sufficient experience of operating equipment imported from abroad are able to export the capability to organize, produce and sell in foreign locations by the medium of direct investments, the main examples being Hong Kong and Singapore.[2] In construction and other service sectors, countries which have built up the relevant skills and infrastructure, and which possess the right blend of cost advantages and government support, have shown themselves capable of winning substantial foreign contracts and exporting their know-how: Korea, Taiwan, India, Pakistan and Egypt in Asia and Africa, and Argentina, Mexico and Brazil in Latin America, are particularly active in these spheres. Exports of technology appear to be among the most dynamic elements of these countries' manufactured exports, and they are directed not only at other less-developed countries but even, in some cases, at the advanced industrial countries.

This bald statement often provokes surprise and disbelief in economists who are concerned with developing countries but are not intimately in touch with their industrial progress. Why should it? Surely, with growing industrial maturity, the industrialized developing countries must exhibit some of the forms of comparative advantage that characterize the advanced countries, despite being backward in several other important respects. Perhaps economists are insufficiently aware of history. Perhaps they are trapped by the static framework in which they analyze the world. Certainly, the various branches of economics that may be concerned with it are not properly equipped to deal with the new phenomenon.

First, the general literature on the nature of technical progress, heavily influenced by Schumpeter, has tended to focus on discrete jumps in technology as the major source of technical progress. This 'breakthrough syndrome' has detracted from a proper appreciation of minor changes made in the process of diffusion, imitation and adaptation: such 'minor' activity is often a greater source of productivity gains than the spectacular jumps in technology.[3] Thus, the likelihood that developing countries are 'innovators' has generally escaped notice, since they are clearly not active on the frontiers of major innovations. Thus, even as perceptive a student of technological evolution as Rosenberg, who argues strongly for a proper appreciation of 'minor' innovation, notes in a recent publication:

> Many of the major innovations in Western technology have emerged in the capital goods sector of the economy. But underdeveloped countries with little or no organized domestic capital goods sector simply have not had the opportunity to make capital saving innovations because they have not had the capital goods industry necessary for them. Under these circumstances, such countries have typically imported their capital goods from abroad, but this had meant that *they have not developed the technological base of skills, knowledge, facilities and organization upon which further technical progress so largely depends*.[4]

Rosenberg's remark is more by way of casual observation rather than the result of an empirical investigation, but it reflects a widely held, and understandable, belief that poor countries, with little scientific experience or infrastructure, insignificant investments in research and relatively small and new local enterprises, are unlikely to be able to contribute much in terms of technical progress. They are even less likely to be able to compete with established giants of the developed world in international technology markets. The possibility that they are achieving significant innovation within the confines of their own industrial experience, even without massive research expenditure, and that this innovation is exportable, is naturally not taken seriously into account.

Second, standard economic theory has also not been particularly helpful in preparing the ground for accepting such a possibility. 'Pure' neo-classical theory has always found it difficult to deal with technical progress in a realistic way. It has always been bedevilled, particularly at the micro-economic level, by trying to distinguish moves along a universally known production function (factor substitution in response to relative price changes) from moves of the function itself (exogenously provided technical progress).[5] Empirical investigations have been further complicated by the effects of scale, 'X-efficiency' and learning-by-doing, and the best that has been achieved is the depiction of technical progress as the residual left after the influence of factor changes has been taken into account. In recent years, however, it is increasingly accepted that, from the firm's point of view, the usual neo-classical portrayal of innovation was misleading, and that firms did not operate on a well-understood production function. They groped in an uncertain and half-understood environment, from a point (rather than the entire frontier) which may or may not be on the frontier of knowledge, and that any move from the point involved a risky and costly search.[6] This characterization of technology, in which *any* product or process discovered which is new to the firm concerned is an 'innovation', is much more amenable to application in developing countries than the neo-classical model. We shall return to this 'evolutionary' view of innovation in later chapters. Here we should only note that the sorts of technical activity being undertaken in developing countries may well be within production possibility frontiers and may well be to substitute factors rather than to increase their overall productivity: but it is 'innovation' nonetheless, worthy of study because it yields the innovator a unique advantage in international markets to sell a process or a product of which only he has operating knowledge.[7] The study of such innovation would be very difficult to conceptualize under the usual theoretical constructs of neo-classical economics.

Third, despite the considerable interest which has been aroused by the rapid and sustained growth of manufactured exports by the more advanced developing economies, the fact that they are diversifying into the most skill-intensive of all exports – technology itself – has somehow escaped analytical notice.[8] In part this is just the normal academic lag behind developments in the real world. In part, however, it is due to the role that trade theory assigns to developing countries. Neo-classical models of comparative advantage of the Heckscher–Ohlin variety predict that their exports will consist of labour-intensive products. More recent theories, which discard the simple division of factors of production into labour and capital and concentrate instead on such advantages as scale economies, skills, technology and product differentiation, assign to developing countries the role of exporting low-skill, low-technology and non-scale-economy-intensive products.[9] Both sets of theories do not, in consequence, prepare us for the proposition that even very poor countries may develop a comparative advantage in the sales of skills and know-how themselves. To the extent that observers have acknowledged the rapid diversification of Third World manufactured exports, they have traced its cause to the entry of foreign skills and technology *via* the activities of export-orientated multinational companies or foreign buying groups. Local technical efforts, if acknowledged at all, have been characterized as factor substituting (i.e. using more labour intensive, simpler processes) rather than as 'innovative' in any sense. The possibility that sufficient 'human capital' has been accumulated to the extent that skill intensive exports of developed countries can be challenged is generally not taken into account.

Fourth, there *is* a large literature on technology in developing countries, with two different strands to it, but it also ignores, or plays down, the possibility that developing countries are innovating in large sections of modern industry and exporting their innovations. The first strand, concerned with the problem of technical choice and, more recently, with 'appropriate' technology, has either not dealt with the problem of technical change, or else has come to rather pessimistic conclusions about the possibilities of technical adaptation in existing modern manufacturing industry.[10] The other strand, concerned with the problem of technology transfer, has treated developing countries as passive recipients of technology from the advanced countries, and concentrated on the imperfections of the technology market that raise the costs of buying technology to the hapless developing countries.[11] Much of the recent rhetoric surrounding the subject of technology transfer has, therefore, found it convenient to ignore the fact that developing countries are becoming significant generators and exporters of industrial technology.

In sum, there are several reasons why the current intellectual apparatus with which we view technology and developing countries biases us against looking for innovation in those countries. If developing countries are, indeed, innovating and exporting technology, many of the attitudes which have become ingrained in us may have to be modified. In particular, it may call for a more realistic assessment of the process of 'evolutionary' technical change at the enterprise level in developing countries, discarding the general presumption that they have little or no capacity for independent technical progress. It may call for a rethinking of the modern conventional wisdom on comparative advantage, which assigns to developing countries a role at the bottom of the skill and technology ladder. And it will certainly challenge the usual portrayal of the international technology market as one where developing countries are only the passive, and generally hard-done-by, recipients of technology from the industrialized world.

It will be argued below that developing countries are in fact 'innovating' significantly in the evolutionary sense of the term. Their innovations may not constitute major breakthroughs at the frontiers of advanced technology, but contribute nonetheless to their productive and competitive abilities. Their revealed comparative advantage in exporting a wide range of different technologies to a broad set of countries suggests that their dynamic comparative advantage lies in certain sorts of high-skill activities not normally associated with poor countries. Moreover, it shows that technology flows are not entirely uni-directional: for the technologies in which developing countries have asserted their advantage, we may well see an increasing flow of exports to the developed countries as well as the increasing collaboration between developed and developing countries in selling technology. This is not to suggest that developing countries can match developed ones over the whole spectrum of technology generation: clearly they cannot, but their frontiers lie well beyond what is normally believed.

The evidence adduced in this book also suggests strongly that different developing countries have markedly different patterns of revealed comparative advantage in technology exports. To some extent this may be traced to their relative sizes and educational structures, industrialization and technology policies. It has now become a commonplace that protectionist import substitution strategies have not been very successful and have fostered industrial structures that are inefficient, undynamic and uncompetitive. With the general enthusiasm for liberal outward-looking strategies has gone an espousal of *laissez-faire* economic policy on all international transactions. The evidence on technology exports calls for some re-appraisal of this current orthodoxy. While over-enthusiastic protectionism may well have created some industries which will never be internationally viable, the protection of certain activities has also led to the emergence of competitive and technologically dynamic enterprises in range of complex manufacturing industries.

It is not that the theoretical case for or against protection is affected: merely that the extent of desirable protection on 'infant industry' grounds is probably much larger than most present thinking has allowed for. Infant industry arguments can extend to the protection of indigenous technological development rather than simply to the protection required to reach a static efficiency frontier. We should try to distinguish analytically between these two forms of protection – of domestic technological effort as distinct from that of simply domestic production with a given (generally foreign) technology – though in practice they may often be closely intertwined; only then should we try to assess whether or not there is a good case for recommending protection for the former. By not separating the two kinds of protection, and by taking a rather static view of the efficiency of protected industries, the current development literature has probably swung too much in favour of liberal policies.

One final point about the implications of technology exports. If a case can be made in favour of protecting domestic technological effort, we may need to look again at the role of the import of foreign technology in the technological development of poor countries. We are not concerned here about the appropriateness of the technology, but only with its effect on local capabilities; we accept also that the developed countries, and the multinational companies based in them, are in the forefront of major technological change. Is the transmission of technology via the MNC, efficient as it is, necessarily beneficial to the host country? Is the role of the MNC the positive one (as Findlay, 1978, suggests on theoretical grounds) of acting as a direct conduit for advanced technology, the catalyst in the process of latecomers catching up with the starters? Or is this undoubtedly positive aspect of their operations mixed with a partly negative one (as Lall and Streeten, 1977, argue) of inhibiting local technological development? This book argues the latter point of view, but with several qualifications: the effect of easy access to foreign technology may be inhibiting whether it occurs through MNC investment or licensing; this effect should diminish as the technological base of the country grows; and MNCs can in certain conditions contribute significantly to local technological capabilities. These are important matters about which there is much conjecture but little hard evidence – that gleaned from technology exports permits some tentative inferences to be drawn.

There are then several reasons why the study of technology exports is an intellectually challenging and practically significant task. The themes mentioned above will recur through the book, though the paucity of understanding of the micro-economics of technical progress prevents us from arriving at more than rather speculative conclusions.

The book is laid out as follows. The next chapter describes what is meant here by 'technology exports'. The following six chapters, comprising Part B, review the evidence on such exports by India. They also present the available evidence on comparable exports by other developing countries: limited as it is, the evidence shows interesting contrasts between the different exporters. Part C of the book provides an assessment of the picture conveyed in the previous part and tries to account for the differences between the revealed comparative advantage of the different countries. A set of Appendix tables furnishes the sketchy information that is at hand about Indian technology exports.

2 The Definition of 'Technology Exports'

There is no commonly accepted definition of 'technology exports'. In a very broad sense the term could be taken to comprise the export of all goods and services that embody some element of local productive knowledge, but this is too general a definition for our purposes. The literature on international transfer of technology utilizes a definition which comes close to our requirement. It is based on the viewpoint of the buyer rather than that of the seller, and includes in its compass the provision of such skills, designs, services and patents which contribute directly to the productive capacity of the former. Thus, intermediate and consumer goods imports are not counted as technology purchases (though in some contexts we may conceive of these also adding to the buyer's productive capacity). The import of capital goods, while a form of 'embodied' technology, is also generally excluded under this convention. The line drawn is clearly somewhat arbitrary, but it seems to be based on the distinction between transactions in skills and productive knowledge as such, on the one hand, and in commodities and equipment, on the other.

This particular delineation of the scope of 'technology exports' is convenient for us for two reasons. First, a study of the provision of 'pure' skills focuses attention directly on the segment of the trade spectrum where the human capital and know-how by themselves determine comparative advantage. It is precisely these factors which we wish to examine here in the context of the development of poor countries. Second, the fact that a developing country is exporting technology in this sense is prima-facie evidence that it has accumulated skills and knowledge within its own borders, based upon its own experience and effort. For the export of capital or other goods, this inference is more difficult to justify: the goods exported may simply be assembled, by a foreign affiliate or a local licensee, from imported components, according to designs and specifications provided from abroad. The local skills embodied may, at the minimum, only be very simple ones of soldering or packaging, acquired in a fortnight of training. Even the exports of 'technology' can, however, represent different degrees and forms of local skill, and this will be discussed further below; but there is clearly *some* local skill which has reached international levels of competitiveness.

Our definition of technology exports is rather fuzzy around the edges. It is not clear if all skills and services exported should be counted as 'technology' in the sense of adding to the productive capacity of the buyer. Should the sale of health-related services, or those for civil construction or tourism, be included? Should the provision of managerial, financial or other skills be counted as 'technology'? The main emphasis of the technology transfer literature has been on skills related to industry and infrastructure, and these constitute the bulk of the value of technical sales and receipts: however, there is no strong reason why other activities should be excluded. Perhaps fortunately, the paucity of available information renders this problem a rather academic one for present purposes. The greatest amount of data available is in fact on industrial technology transactions and, somewhat less so, for civil construction activity. Other sorts of exports, while they certainly exist, are so poorly documented that we can only make passing reference to them.

Within the industrial sector, however, a variety of technical (in the narrow meaning) and other (managerial, financial, marketing) skills are being exported, and the data do not allow us to differentiate between them meaningfully. For analytical purposes the distinction between the various types of skills is very important, and we shall discuss it as far as the evidence permits, but no rigorous quantitative differentiation can be attempted.

There is another related set of problems. The export of technical skills and services is sometimes undertaken on its own, but it is often part of a larger package of the provision of capital goods and equipment. The sale of sophisticated machines usually requires considerable direct contact between the two sides of the transaction: before the actual sale to determine the requirements, establish precise specification, submit and negotiate tenders, arrange for credit and for complementary services; and after the sale, to install the equipment, service it, provide spares and sometimes the requisite training. The provision of these pre- and after-sales services is a necessary part of selling capital goods, and it clearly requires highly-developed skills, but should it be counted as the export of 'technology'? Probably not, if we are excluding the sale of capital goods as such from our definition.

But take the more complex case of the sale, not of individual pieces of equipment, but of entire 'turnkey' plants. Here the seller provides a whole, or part of a complex of, services (discussed below) from the evaluation of the project, its basic design, detailed engineering, purchase of equipment, construction of plant and infrastructure, commissioning, training and even, in some cases, marketing assistance. There are clearly strong reasons for counting the provision of the various services as a technology export, even if the provision of capital goods is an integral part of the transaction.

Or take the case of direct investment overseas. Here the technology exporter provides, not only some or all of the technical skills involved in a turnkey project, but also continuous managerial, technical and marketing services after the plant is operational. In several cases the investor also provides some or all of the capital goods required and some or all of the intermediate goods required for the manufacture of the final product. Clearly, there is an export of 'technology' involved, but, equally clearly, it is well nigh impossible to separate the technical component of the whole package.

We have to bear these complexities in mind when considering the evidence on technology exports. A final point before we categorize these exports in detail: we only include the *sales of technology that take place under market conditions*, i.e. the emigration of skilled labour and the provision of government technical assistance are excluded. While such other provisions of technology undoubtedly embody skill accumulation in the exporting country, we wish to focus on transactions which are known to be internationally competitive. We realize that 'open' markets for technology sales and purchases are riddled with all sorts of imperfections: buyers may be poorly informed, biased or subject to political considerations; prices are not uniform, and are difficult to establish for particular elements of large packages; bargaining and such factors as 'kickbacks' may figure in the negotiation of contracts; and so on. Despite these factors, these markets are competitive to a greater or lesser extent, and the ability to penetrate them says something about the abilities of the developing countries. In fact, to the extent that political considerations, inherited biases and lack of information work in favour of established technology sellers from powerful developed countries, the new developing-country entrants probably reveal an even greater competitive ability in winning contracts than would be the case under more 'perfect' markets.

We may group technology exports into three broad groups: industrial, civil construction and other services. Each of these can entail the export of some embodied technology (capital goods) along with the sale of skills and know-how. These constitute the subject of our study. To take them in turn:

1. *Industrial*: Industrial technology exports may be defined broadly to include, besides manufacturing activity, such sectors as power generation and distribution, desalination, sewage, transportation (but excluding road or airport construction) and communications. The line between civil and industrial construction is sometimes difficult to draw, especially when the same firms undertake both sorts of activity, so some arbitrary judgment is

necessarily involved. Industrial technology exports cover four different types of activity, ranging from the most embodied to the most disembodied, viz.

(a) *turnkey* projects, which are partially embodied (to the extent that they entail exports of local equipment) and partially disembodied (to the extent that they entail feasibility studies, project design, tendering, construction, commissioning and training);
(b) *direct investments* or joint ventures, which are, in similar fashion, partially embodied and partially disembodied, except that the disembodied managerial and financial relationship is of a continuing nature, and may also involve later exports of capital equipment for expansion;
(c) *licensing*, which, broadly defined, is mostly disembodied (in the form of technical services, patents, know-how, designs) but may have an embodied element if the licensor provides equipment to manufacture or assemble a product abroad; and
(d) *consultancy* (engineering and managerial), which is wholly disembodied, though it may, and usually does, lead to the export of technology in other forms by other agents.

2. *Civil construction*: Here also technology exports may take a variety of forms, but in general they contain a small element of embodied exports and a large element of the provision of engineering, technical and organizational services, occasionally accompanied by the temporary export of semi- or unskilled labour.
3. *Other services*: The export of 'other' technical services, e.g. hotel management or ownership, financial services, health services, agricultural services, and the like, are more or less wholly disembodied. As with industrial consultancy, however, they may (though less often) result in the export of equipment from the home country.

This 'product-wise' classification of technology exports is still rather general, and conceals some important information about precisely what the technology exporter is selling. Two different countries, for instance, involved in turnkey projects abroad may be performing completely different ranges of functions, with significantly different implications for the problem we are investigating. One may be providing (through a number of collaborating enterprises) the whole range of services from preliminary studies, through basic and detailed engineering and the construction of capital goods to final commissioning and training. The other may only be the prime contractor which has won the contract, and which then buys the basic process technology from one developed country, the capital goods from another, carries out the construction itself and buys training services elsewhere. The *local technological content* of the work performed is completely different in the two cases: in the first, the exporting country has the ability to reproduce the entire technology itself and to organize its transmission abroad; in the second, the exporter only manifests organizational ability and the ability to carry out the civil construction part of the job. Both are selling skills – but it is clearly necessary to distinguish the form which this takes.

For turnkey project work a useful classification of the enterprises involved would be as follows:[1]

prime contractors;
capital goods producers;
process developers;
specialized consultants;
specialized contractors;
other agencies (banks, traders, government, etc.).

One or more of the functions performed by these enterprises could be merged if the same firm acts, say, as the prime contractor and capital goods producer, but normally there will be different firms, each with its own speciality, involved. In certain cases the prime contractor could be a government 'lead' agency set up specifically to bid for contractors. Engineering firms sometimes also develop production processes, and so are able to do the basic technical design and development work, but in general they are not capable of providing process know-how for a wide range of industries.

Unfortunately, our data do not allow us to draw all the fine distinctions that we would like to in reviewing the evidence. We can usually identify the main specialisation of the technology exporters, but, because of the diversity of the roles they can perform, we cannot always identify the precise content of their foreign activity. We do, however, attempt to distinguish in a rough way the local technological content of the exports.

The main focus of this study is on industrial technology exports in the four forms described above. This is, as noted earlier, dictated by the nature of information available rather than a belief that they are more economically significant. Because the exports of capital goods are an integral part of some important forms of technology exports, some attention is also given at the start of the next part to the Indian performance in this area. This furnishes a useful backdrop to the discussion of technology exports proper. Let us turn now to the evidence.

Part B

The Indian Experience

3 Some Background Information

I INTRODUCTION

This part of the study presents the evidence collected by the author on Indian technology exports. The sources are diverse: newspapers, magazines, personal communication and some published data. With the exception of capital goods exports, the figures given are incomplete. The greatest gaps are probably in purely disembodied industrial exports like consultancy and licensing and in non-industrial technology exports. The direct investment information, while quantitatively not satisfactory, is reasonably complete in terms of the identity, destination and activity of the enterprises. Turnkey project data are patchy, though it is believed that most of the important exports in the non-traditional sectors have been mentioned; a number of smaller projects and projects in traditional sectors (i.e. textiles, cement, food processing) have not, however, been captured. Despite these deficiencies, these are the first data to be collected and published on this subject, and must serve until more comprehensive field work is conducted.

Before we come to technology exports, as defined here, it may be useful to review the evidence on recent exports of capital goods by India for two reasons: first, capital goods are, as noted earlier, intimately related to disembodied technology exports, and it is difficult to extricate them from the available data on technology transactions. Second, it may be interesting, for the same reason, to compare the Indian performance in exporting capital goods to that of other technology exporting developing countries. Does success in exporting one imply success in exporting the other? How do the two such closely related products relate to each other? Do different institutional arrangements for exporting capital goods (in particular, the different roles permitted to foreign as compared to local enterprises) affect relative performance in exporting technology? These are important questions to which we shall suggest some tentative answers later in the work. Let us first consider the evidence on capital goods exports.

II CAPITAL GOODS EXPORTS

The Engineering Export Promotion Council (EEPC) of India provides fairly complete and detailed data, on an annual basis, of Indian exports of engineering goods, and the following data are taken from its 1976/77 report. In the nature of the goods, it is very difficult to separate capital goods from other types of engineering goods exports. Clearly, some types of engineering consumer goods are not usable in the production of other goods (passenger cars and bicycles) and so are not capital goods in any sense, but the definition of 'consumer durables' adopted by the EEPC is rather wide, and includes such items as diesel engines and pumps, data processing machines and small tools. It also employs a category 'Primarily steel and pig-iron items' which is a pot-pourri of different types of consumer, intermediate and capital goods.[1] It would have been extremely tedious, and not particularly revealing, to rearrange and analyze all the detailed data for a number of years. What we have done here is first to present information on all engineering exports together, then on these exports reclassified by more meaningful groupings at a fairly aggregate level, and then on industrial plant and machinery narrowly defined.

Total Indian engineering exports have grown from Rs. 51.6 million ($10 .7 m) in 1956/7 to Rs. 103.1 million ($21.5 m) in 1960/1, Rs. 294.1 million ($39.2 m) in 1965/6 Rs. 1157.7 million ($154.2 m) in 1970/71, Rs. 4082.2 million ($390.3 m) in 1975/6 and Rs. 5516.8 million ($535.6 m) in 1976/77.[1] Thus, they have grown over a hundredfold in rupee terms and about fiftyfold in dollar terms over twenty years, an average annual growth rate of 25 per cent in dollars and 53 per cent in rupees. They have been among the most dynamic elements of Indian exports, and have accounted for the largest increment (in constant prices) of Indian exports between 1960/1 and 1974/5 – 19.1 per cent of the total gain in fourteen years, as compared to 16 per cent for iron ore, 14 per cent for handicrafts, 10 per cent for sugar, 8 per cent each for chemicals and cotton textiles and garments (of these garments have grown much faster than textiles), and less for other items.[3] Engineering exports now constitute around 11 per cent of total exports and are the leading category in terms of absolute value. They contribute 19 per cent of total manufactured exports, and press reports subsequent to 1976/7 show that they have maintained a healthy rate of growth in 1977/8 (13 per cent in rupee value) and 1978/9 (precise figures are not available, but preliminary reports suggest that they have increased by 20 per cent in the face of practically stagnant total exports).

Table 3.1 below shows the broad geographical distribution of engineering exports. Table A.1 gives the destinations in much greater detail, showing

TABLE 3.1 Engineering goods exports. % Composition according to major destination

	Export in 1956/7 In crore[a] Rs.	Export in 1956/7 % Composition	Export in 1976/77 In crore[a] Rs.	Export in 1976/77 % Composition
1. Asia	3.76	72.86	315.82	57.24
South East Asia	1.77	34.28	117.18	21.24
West Asia	1.99	38.58	198.64	36.00
2. Africa	1.20	23.23	79.28	14.37
3. Europe	0.01	0.21	105.58	19.68
East Europe	—	—	37.86	6.86
West Europe	0.01	0.21	70.72	12.82
4. America incl. Caribbean Islands	0.02	0.36	36.45	6.61
5. Oceanic Islands	0.10	1.96	4.56	0.83
6. Australasia	0.07	1.38	6.99	1.27
Total	5.16	100.00	551.68	100.00

NOTE

[a] In crores of rupees: one crore equals 10 million.

SOURCE

EEPC, *Handbook of Export Statistics 1976–77* (Calcutta, 1978) p.6.

engineering exports to 57 individual countries for the period 1974/5 to 1976/7 as well as total exports to them. Among the various points of interest in these figures, we may note that (a) the developing world (essentially Asia and Africa, since India exports relatively little to the developing areas of the Americas) accounts for 72 per cent of engineering goods exports in 1976/7, (b) the share of the developed world, while relatively small, had risen from a mere 4 per cent in 1956/7 to 28 per cent by 1976/7, and (c) while a number of developed countries are, as Table A.1 shows, major buyers of Indian engineering goods (in particular, the US, West Germany and the UK, as well as the USSR in the socialist bloc), engineering exports account for small proportions of India's total exports to them, generally much lower than the engineering share (of 11 per cent) in total exports.[4] In contrast, engineering goods account (with a few exceptions) for much higher proportions in exports to other developing countries. It is also worth noting that the proportion of engineering goods in total exports is highly variable over time for developing countries, much more so than for developed ones.

Table 3.2 summarizes the industrial composition of engineering goods exports for 1972/3 to 1976/7; it is derived from more detailed data presented

TABLE 3.2 Industrial composition of engineering goods exports, 1972/3 to 1976/7

Item	*1972/3 value(%)*		*1973/4 value(%)*		*1974/5 value(%)*		*1975/6 value(%)*		*1976/7 value (%)*		% *Growth over period*
Industrial plant	114.6	(8.1)	159.6	(8.2)	312.3	(8.7)	455.3	(11.2)	431.9	(7.8)	276.9
Machinery, tools	203.2	(14.4)	280.7	(14.5)	531.6	(15.2)	673.1	(16.5)	910.4	(16.5)	348.0
Electrical machinery	345.2	(24.5)	379.4	(19.6)	664.6	(19.0)	826.6	(20.2)	962.4	(17.4)	178.8
Transport equipment	289.4	(20.5)	386.7	(20.0)	647.0	(18.5)	890.5	(21.8)	939.1	(17.0)	224.5
Metal products	386.7	(27.4)	627.7	(32.4)	1170.2	(33.5)	1057.4	(25.9)	2036.7	(36.9)	422.7
Other	71.7	(5.1)	100.6	(5.2)	165.4	(4.7)	178.9	(4.3)	236.0	(4.3)	229.1
Total	1410.8	(100)	1934.7	(100)	3391.1	(100)	4082.2	(100)	5516.8	(100)	291.0

NOTE
Percentages may not sum to 100 because of rounding.
SOURCE
See Table A.2.

in Table A.2. The industries shown here have been rearranged from the (unsatisfactory) classification used by the EEPC. Both tables show overall growth rates for the four-year period. Engineering goods as a whole grew at 73 per cent per annum on average (uncompounded and in current rupees), a healthy performance in comparison to the fact that in 1968–72 they had grown at 24.4 per cent on average.

In broad sectoral terms, metal products have dominated engineering goods exports in terms both of absolute values as well as rates of growth. They have increased their share from 27 to 37 per cent over the period, and exhibited annual average rates of growth of over 100 per cent. Within this group, pipes and tubes, wire products, sanitary ware and non-ferrous products have been particularly important, all enjoying growth rates well above the average. Bright bars, ferrous hollow-ware and industrial fasteners have also grown rapidly, but are smaller in terms of absolute values. Since metal products are not capital goods in the usual meaning of the term, however, we should discount for this item: without it, total engineering products grew by 240 per cent over the period, or an average of 60 per cent per annum.

The next fastest growing category of engineering exports – though the fourth in terms of 1976/7 values – is machinery and tools, which have grown at some 83 per cent per annum on average. The fastest growing products here are pumps and machine tools, though diesel engines and compressors and hand tools (which grew at around average rates) were the main earners of foreign exchange in absolute terms.

Electrical machinery has grown slower than average, at around 45 per cent per annum (uncompounded), and has steadily lost its share of total engineering exports. Within this group, however, fans and 'other' electrical products (lamps, appliances, light consumer goods) have risen faster than average. The largest exchange earner, wires and cables, has only grown at 24 per cent per year on average. The two high technology products in the group – power machinery and switchgear, and electronics – have grown at 60 per cent and 42 per cent respectively. The power machinery performance is quite respectable, in view of the complexity and capital intensity of the industry; that of electronics is, however, rather poor, in view of the staggering rates of expansion of this sector in some South East Asian and Latin American economies.

Industrial plant has grown on average at 69 per cent per annum, with above average expansion in textile and jute plant (a setback in 1976/7 may in fact understate its longer term expansion potential), food processing plant and 'other' plant. The last item, containing a variety of equipment considered in more detail below, has grown at 117 per cent per annum, and is the major contributor to exports in this group, accounting for nearly half the total.

Transport equipment has grown at 56 per cent per annum, with the major contribution coming from complete commercial vehicles (including jeeps) and vehicle parts. Taken together, automotive products totalled Rs. 547 million in 1976/7, second only to pipes and tubes (Rs. 587 million) among all engineering goods exports. Ships have grown much faster (at nearly 400 per cent per annum) but from a very small base. Bicycles and parts, earlier the largest export in this group, have grown at around 25 per cent per annum and have been rapidly overtaken by automotive products.

Among 'other' engineering goods, the overall expansion has been slower than average, with only miscellaneous goods beating the average. Sewing and knitting machines have stagnated. Heating and cooling equipment, and instruments of various types, have demonstrated reasonable rates of expansion.

It should be clear why it is difficult to separate 'capital goods' proper from engineering goods in general, and why we have left the groups together in this section. Since the main focus of our study is on industrial machinery and plant, details of electrical and non-electrical machinery exports have been extracted from the EEPC report and presented in Table A.3. This shows the wide diversity attained by Indian machinery manufacturers in their foreign sales, and also the great variety of experience of individual items in terms of growth and stability. A great deal of the jumps and falls in plant exports are due, of course, to their 'lumpiness' and the fact that many individual items are not sufficiently large or geographically dispersed to be able to smooth out the effect of particular sales.

A few comments on Indian engineering exports may be in order:

First, the nature of the competitive edge that they enjoy in foreign markets. While the data are not available to demonstrate this statistically,[5] sufficient impressionistic and anecdotal evidence exists that sales to the developed countries are of simpler, more standardized and less skill-intensive products than those to other developing countries. In developed industrial markets, the main strength of Indian engineering products is price competitiveness for relatively low-technology items. In developing markets, on the other hand, it is their relatively higher levels of skill combined with greater appropriateness to local conditions: Indian machines are more complex than the importing countries could make themselves, but they are not as sophisticated, as large, as expensive or as automated as machines that developed countries would supply. This is in keeping with India's place between the highly industrialized and the unindustrialized countries: its experience of industrialization seems to have given it a dynamic comparative advantage in certain intermediate types of capital goods manufacture, for which the technology has been mastered and adapted.[6]

The key word, however, is 'dynamic'. Even in exports to highly developed countries, modest breakthroughs have been made in the export of locally designed high-technology items like numerically-controlled machine tools, advanced industrial boilers, telecommunication equipment and computer memories.[7] These products may not be on the forefront of innovation, but they represent a far higher level of technical achievement than 'traditional' engineering products like small tools, wires, cutlery and pipes. Of course, it may be a long time before India can capture a large enough market in developed countries for high-skill items to alter the overall pattern described above, and it may not in the foreseeable future develop a competitive advantage in commodities which require major research effort merely to keep up with changes in technology. The 'product cycle' model applies rather nicely to Indian exports: a fact to which we shall return later in the book.

Second, whether changes in the pattern of Indian engineering exports are due to government subsidy (and other export-promotion policies) or domestic excess capacity rather than to genuine changes in local competitiveness cannot be judged from this data. The evaluation of the net effect of protective and subsidization measures is a complex task (see Bhagwati and Srinivasan, 1975) which lies outside the immediate purpose (and resources) of this author, but it seems reasonable to argue that both a shift to more outward-looking policies and the accumulation of local skills and technology were responsible for success in selling complex items of equipment abroad. Domestic excess capacity probably helped in this by forcing firms to look for new customers and to offer low prices (this is of course the normal practice for all exporters breaking into new markets), but the sustained rise in engineering exports over a fairly long period, both boom and recession, suggests that this was not the only factor responsible. In any case, since buyers of engineering products look more closely at performance, reliability and technical characteristics than at price *per se*, the growing success of Indian machinery does indicate that on this score they had achieved international competitiveness.

On this point we should note that analysts like Frankena (1973–4) and others made much, some 5–7 years ago, of design deficiencies in Indian engineering exports and the lack of domestic design, research and development work. On present evidence, however, a few qualification must be made to their argument. (i) A number of engineering goods producers, particularly large ones, have set up their own R & D facilities in recent years and have launched efforts to innovate on basic design. (ii) Even in the absence of explicit R & D efforts it is likely that manufacturers have built up sufficient technological capability from production, trouble-shooting and 'minor' technical work to gain a competitive edge in international markets.[8] (iii) While many Indian machinery exports may be based on old designs they evidently enjoy a market among producers in other developing countries who do not find it economical to purchase the latest equipment.[9] Furthermore, many of these old designs have been modified, strengthened and adapted to use different materials;[10] in view of their recent success abroad it may be safely inferred that such adaptations outweighed their disadvantages due to old design.[11] While a number of manufacturers may still lack adequate design capabilities, it would be unwise to take a static view of this aspect of production: the evidence adduced later will provide further support for a more dynamic and optimistic view.

Third, we must be careful, on the other hand, not to overstress the competitive abilities of developing country engineering producers like India. There are areas of engineering technology where they are competitive, either because they are fairly near a relatively slow-moving technological frontier, or because in some activities even a long lag does not prevent them from finding markets in lesser-developed countries. There are other, and large, areas of technology where the frontiers are moving too rapidly for them to catch up, or where even a small lag leads to products which are unsaleable (i.e. inefficient at any reasonable configuration of factor prices). Advanced electronics or highly sophisticated transfer-line equipment are good examples of products where developing countries cannot invest enough in the technological assimilation process to become internationally competitive. Here the only way to enter export markets may be to subcontract certain labour intensive processes (which are relatively less skill- and technology-intensive) from multinational companies from the advanced countries, at least for items where transport and training costs render such subcontracting economically feasible.[12] India has been very tardy in entering this form of export activity, and, as we shall see in a moment, this has involved a substantial loss of export earnings in comparison to some other developing countries.

Finally, the general role of multinationals (MNCs) from developed countries in Indian exports. It is generally acknowledged that foreign enterprises contribute relatively little to Indian manufactured exports.[13] Even in the high-skill, high-technology sector of engineering goods, where most developing countries depend heavily on MNCs, only 14 per cent of Indian exports are accounted for by foreign-controlled firms[14] (however, the role of joint-ventures and foreign licensees may be important). There is little reason to expect this share to change sharply in the future.

TABLE 3.3 India: Statistics for major engineering sectors, 1974/75

Industry	Factories (no.)	Productive Capital (Rs. million)[a]	Employment (nos '000)	Output (Rs. million)[a]	Capital	
					Employee (Rs. '000)	Output (%)
1. Basic metals and alloys	4215	21,077.2	421.7	26,360.2	50.0	80.0
2. Metal products, parts	4434	1721.7	172.2	6835.4	10.0	25.2
3. Machinery and tools (non-elect.)	4713	4831.5	335.5	14,856.4	14.4	32.5
4. Electrical machinery and parts	2380	4340.1	254.4	13,843.6	17.1	31.4
5. Transport equipment and parts	1600	4027.6	395.2	13,607.4	10.2	29.6
6. Other engineering[b]	551	554.8	32.9	952.7	16.9	58.2
7. Repairs[c]	2047	858.0	111.6	2411.6	7.7	35.6
Total Engineering industry	20,053	37,410.9	1723.4	78,866.3	21.7	47.4
All industry	64,217	179,322.3	6052.8	260,990.1	29.6	68.7
Engineering % of all industry	31.2	20.9	28.5	30.2	—	—

NOTES

[a] In 1974/5 one US dollar was exchanged at approximately 8 Rupees.

[b] Includes manufacture of medical scientific equipment, watches and clocks, photographic and optical goods and musical instruments.

[c] Includes electrical repair, automobile repair, and repair of watches, clocks, cycles and motorcycles.

SOURCE

EEPC, *Handbook of Export Statistics, 1976–77* (Calcutta, 1978) p. 29.

Let us now look briefly at the domestic engineering goods sector in the country. Table 3.3 presents some data outlining the size, employment and capital-employed in engineering goods production and its constituent sub-sectors in 1974/75. It shows that the industry comprised some 20,000 establishments (including 2000 for repair), just under one-third of all industrial establishments in India. It employed less capital per employee and per unit of output than industry generally (bearing in mind the usual reservations about census data for capital employed). Within the industry, basic metals and alloys was the largest employer but also the most capital intensive by far (note that basic metals are not counted as engineering products in the export data given above). Metal products are the least capital intensive (ignoring repair work), as we saw earlier, also the major exporter. Transport equipment was slightly more capital intensive, followed by non-electrical machinery, electrical machinery and other engineering. The largest employment per factory was afforded by transport equipment, the least by metal products and parts.

The size and diversity of the domestic engineering industry is worth noting from the point of view of the potential that exists for local 'learning' and internal linkages. There are well over 1 million employees in engineering goods manufacture in India, even if we exclude basic metals, of whom a substantial proportion are fully trained technicians and engineers. These provide the seed-bed for learning by various types of experience (discussed at greater length below), partly by simply tackling ordinary manufacturing problems, partly by engaging in the assimilation and adaptation of imported technology and partly (though still insufficiently) by engaging in basic design and development activity. The industry is large enough to provide a base for engineering consultancy activity and to act as a channel for the transmission of new technology across different user industries. It does not as yet undertake substantial R & D (data on R & D in India are given in Chapter 10 below) and is still relatively unspecialized,[15] but even at given levels of activity it enjoys considerable prospects for building up an export capability for both its products and its know-how.

Let us now quickly look at the record of other developing countries in this field. While the performance of India is respectable it does not match that of other semi-industrial countries which have adopted an aggressive export-orientated policy and invited greater participation by multinational companies. The latest figures provided by the UN (1978) and summarized in Table 3.4 show, for instance, that India exports significantly less than a number of other developing countries and has experienced a slower rate of growth.[16] It also shows the small share of world engineering trade accounted

TABLE 3.4 Exports of engineering products by selected developing countries, 1976 (US $ million and %)

Country	*Total (%)*		*Non-electric machinery (%)*		*Electrical machinery (%)*		*Transport equipment (%)*		*Average annual growth rate of total (1970–6)*
India	310.2	(0.12)	121.8	(0.11)	84.2	(0.15)	104.2	(0.10)	21.8
Argentina	406.3	(0.15)	182.9	(0.17)	37.0	(0.07)	186.5	(0.18)	35.2
Brazil	957.0	(0.35)	389.8	(0.36)	202.7	(0.36)	364.6	(0.36)	46.4
Hong Kong	982.1	(0.36)	111.0	(0.10)	860.5	(1.55)	10.6	(0.01)	26.5
South Korea	1267.8	(0.47)	126.5	(0.12)	803.2	(1.44)	338.1	(0.33)	66.3
Singapore	1662.7	(0.61)	398.2	(0.37)	885.2	(1.59)	379.3	(0.37)	46.2
Yugoslavia	1362.9	(0.50)	405.8	(0.37)	395.6	(0.71)	561.4	(0.55)	23.7
Sub-total	6949.0	(2.57)	1736.0	(1.60)	3268.4	(5.9)	1944.7	(1.90)	—
Other countries	263,607.4	(97.4)	106,930.8	(98.4)	52,399.4	(94.1)	100,464.5	(98.10)	—
World total[a]	270,556.4	(100)	108,666.8	(100)	55,667.8	(100)	102,409.2	(100)	20.3

NOTE
[a] For 35 main exporters (excluding, in the Third World, Mexico and Taiwan)
SOURCE
UN (1978), Tables 1 and 5.

for by the Third World: there is a long way to go before it makes a noticeable dent in this industry.

Some of the export figures for entrepôt centres like Hong Kong and Singapore need to be reduced for re-exported products. A large part of electrical equipment exports from the three far Eastern countries comes from 'offshore assembly' of electronic equipment undertaken by MNCs; such activity has not taken hold in other countries in the table. For Brazil and Argentina (especially the former), however, MNCs play a predominant role in exports of other sorts of machinery. Yugoslavian exports benefit both from special trading relations with Eastern Europe as well as from buy-back arrangments with Western firms; Yugoslavia is now a major 'offshore assembly' centre for German enterprises. To a large extent, therefore, the relatively poor Indian performance in the export of engineering products may be due to its policy of 'going-it-alone', restricting the inflow of foreign capital and technology and making it difficult for offshore assembly activities to take hold in the few bonded zones that have been established.

This is not, however, the whole story. A significant contributor, as many commentators (especially Bhagwati and Srinivasan, 1975) have noted, has also been the inward-looking nature of Indian economic policy, which has consistently and in many subtle ways reduced the incentive to export. While most developing countries that had started with import-substituting policies had, by the early 1970s, switched emphatically to export-promoting ones, Indian policies continued (and still continue) to render the domestic market more attractive than foreign ones. Various licensing, monopoly control and small-industry promotion policies have probably also held back the growth of the largest firms which tend to be the most efficient exporters,[17] although the government is now encouraging the setting up of large trading companies with overseas branches. And, finally, various supply shortages (power, steel), transport bottlenecks and labour problems have contributed to these factors in restraining a sustained attack on foreign markets. The role of political uncertainty is more difficult to determine in the sense that poor prospects at home may induce firms to look abroad; however, to the extent that uncertainty has held back investment, the effect has probably been to reduce exports.

The Indian performance thus reflects in part the constraints of an overall inward-looking policy, in part the effort to 'go-it-alone' and to hold down the growth of large private enterprises, and in part poor planning and other problems. This being said, however, two facts should be stressed: First, in view of the constraints placed by economic policy, the achievement of the *indigenous* engineering industry is all the more creditable. Countries like Brazil, Mexico, South Korea, Hong Kong and Singapore have depended heavily on foreign enterprises to promote the export of sophisticated engineering products.[18] India has relied mainly on domestic enterprises: while its total exports may be smaller, therefore, they may represent a greater deal of technological 'learning' on the part of Indian producers. Second, and more interestingly, the policy of 'going-it-alone' in terms of domestic production and technology may have led to the building up of an indigenous technological capability which is not fully reflected in the exports of engineering products, but which shows itself in a comparative advantage in exporting 'disembodied' forms of technology. The evidence presented in the remainder of this book suggests that this is in fact the case: let us now consider this evidence.

4 Turnkey Plant Exports

The demand for turnkey industrial plant[1] generally has risen rapidly in the last two decades as a growing number of developing countries, without a domestic capital goods base and associated engineering skills, have sought to set up manufacturing industries. The oil-rich Middle Eastern countries have probably led the way, but a large number of others in Africa, Latin America and South East Asia (and now, on a massive scale, China), have sought to import this form of partly embodied and partly disembodied technology. This tendency has been reinforced by the growing desire of many developing countries to retain national ownership of their industry. Turnkey projects are an alternative to direct investment by foreign enterprises, yet they retain the benefit of transferring modern production technology and equipment.

The desire of countries to go for entire plants has induced equipment manufacturers to graduate from the export of particular items in which they specialise to the export of projects which involve the cooperation of several different enterprises (producers of different types of equipment, engineering consultants, process-technology holders, civil constructors, and so on) and financial, as well as government, bodies. Certain Indian equipment exporters have (along with those from other countries) thus found that they had to undertake this broader range of activity in order to export at all.[2] Some have even found that certain buyers, especially in the Middle East, prefer an element of equity participation in the plant sale, so that equipment suppliers have taken the further step of becoming direct investors: a sequence from 'unpackaged' to 'packaged' sales of technology, the reverse of the usual mode of technology transfer from developed countries.[3] This trend has even affected some public sector enterprises, which have not otherwise been eager to enter the direct investment arena.

As we pointed out earlier, data on foreign turnkey activity by Indian firms are far from complete, especially as far as 'traditional' industries (textiles, sugar, cement) are concerned. Here turnkey works have been started a fairly long time ago, and the scales involved were fairly small. Indications exist that our information net has missed out many such projects, but we have not been able to track them down with any precision. For other industries, however, our information coverage is better. The projects involved have tended to be larger, and the majority of them have been undertaken in the 1970s: the sources used, while not systematic in their review, have probably captured most of the important contracts. The amount of detailed information obtained is, however, variable and far from satisfactory. In several cases only the name of the enterprise, its activity and the destination of exports are given. The status of the job (i.e. whether under negotiation, under construction or completed) is often not clear, neither is the sum involved. Nothing is known about alternative bids (the savings achieved by the buyer) or about the real resource cost to India.

Despite these handicaps, the information that has been collected reveals an amazing variety of technology exports in this form, both in terms of the activity involved and the regions served. Table A.4 summarizes this information by industry. It deals respectively with traditional, electrical, chemical, telecommunication, steel and mechanical engineering industries. Some 40-odd enterprises, and over 100 projects, ranging from very 'simple' technologies (textiles, transmission towers, small-scale industries) to 'intermediate' level technologies (paints, pharmaceuticals, rubber, conveyors) and to very complex capital intensive and/or skill-intensive technologies (power generation, basic drugs, machine tools, telephone exchanges, steel mills and petrochemicals) have been identified.

The table really speaks for itself. We only need to supplement it with the following comments:

First, almost all the contracts have been won in open tenders, mostly in the face of competition from established technology suppliers in developed countries. The existence of relatively 'well-informed' buyers, who have been able to evaluate bids on the basis of technical specifications and price rather than (mainly) on the basis of an established reputation, has undoubtedly helped Indian enterprises to break into these markets. The repetition of orders for several contractors (e.g. Bharat Heavy Electricals Ltd. in Malaysia and Libya) testifies to the fact that many enterprises have rendered satisfactory service.

Second, with one or two exceptions (like Gammon and De Smet) all the enterprises involved have been locally owned. Of the latter, there have been a large number of private enterprises and a few government ones. Nearly all of them are large firms (by Indian standards) though a few instances are recorded of medium-sized firms banding together in a consortium to undertake turnkey work abroad. Most of them are specialized equipment producers, though there are a few important exceptions like Engineers India Limited and Engineering Projects India Limited, which have both been set up by the government to do general civil and engineering construction work at home and abroad.

Third, some Indian enterprises have acted as sub-contractors to enterprises from other countries (e.g. EIL from Kellogg Corporation in Sri Lanka, EPIL from Mitsubishi in Iraq). This is not, however, as common as may be expected, given the cost advantages of Indian firms in undertaking detailed engineering, project supervision and other 'labour intensive' (rather than technology intensive) parts of turnkey jobs. The government has, in fact, been pushing the idea of joining bids with developed country enterprises in a number of recent statements, and this is a very promising line of growth in those instances where local enterprises do not possess the level of technology required or (if they do) they are not well-known enough to win contracts on their own.[4]

Fourth, nearly all overseas turnkey jobs have been undertaken after considerable experience of similar work at home. It need hardly be pointed out that the sale of complete plants abroad represents a much 'higher' level of technological activity than that of selling individual pieces of equipment. When the exporting country undertakes to supply the whole package, it has to possess the capability to provide everything from pre-project feasibility studies, basic process engineering and detailed designs to construction and training. Every turnkey project has to be individually designed and engineered to specific needs, which vary according to location, raw materials, scales, product range, labour skills and so on; its execution thus must be based on the experience of putting the whole package together at home, of having mastered, not just the production technology, but also the art of organizing the different specialities and enterprises involved. The various technologies involved in a complete plant may have been originally imported, of course, but they are necessarily re-engineered and adapted for utilization in each new plant, and this is a species of 'minor' innovation work which entails cost and effort and requires considerable expertise.

It follows from this that developing countries like India can only build up a competitive advantage in activities which conform to their own 'learning' base. Thus, technologies which are very new and based on large R & D investments in developed countries (or on tightly held engineering information by the leading firms) are out of their reach, as are the technologies requiring such large-scale facilities that no relevant experience has been gained at home. To simplify, the more stable and well-diffused technology, the more will local enterprises find it possible to approach the levels of developed country suppliers. The more dynamic, complex and large-scale the technology, the more difficult will it be for them to find or recreate the conditions in which they can learn it. The dividing line will fall, not by industry, but by process or product *within* each industry. Every industry, even traditional ones, may have certain technologies which are in the innovative phase, generally beyond the reach of developing countries. And even every 'modern' industry may have a range of older, stable, diffused techniques which are capable of being assimilated and reproduced by developing countries.[5] Thus, we should expect to find developing country enterprises competing as technology exporters only up to a *certain range of technical sophistication within each industry*, lagging behind the 'frontiers' of technology to different extents in different industries.

It follows also that technology exporters may continue to be simultaneously importers of technology: Bharat Heavy Electricals Limited (BHEL), the largest single turnkey plant exporter from the country, has recently negotiated a long-term agreement with Siemens to exchange freely all technological information.[6] For BHEL the benefit clearly lies in gaining access to the results of expensive and large-scale R & D beyond its own capabilities, and so developing its product range into larger and more powerful generators. For Siemens, there may also be some technological benefit beyond the royalties received (the details of the arrangement are not available to the present author) in terms of the particular adaptations made by BHEL to render power plants suitable to tropical conditions. A number of other major turnkey plant exporters also rely occasionally on the import of specific items of foreign technology.

Finally, it should be noted that there is no obvious correlation (at least in the incomplete data at hand) between success in turnkey exports and in exports of capital goods. Thus, electrical machinery seems to be flourishing in turnkey projects but growing relatively slowly in relation to engineering products more generally. The category of 'industrial plant' in Table 3.2 above is also relatively non-dynamic, while the impression conveyed by the evidence on turnkey projects is one of fairly rapid growth. However, the two sets of data are not readily comparable. Besides the obvious gaps in turnkey data in terms of annual earnings abroad, the exports of the equipment involved in any given project may span a number of different engineering categories, e.g. an electrical project may involve some 'metal products' and some 'machinery and tools'. As indicated elsewhere, there are signs that exporters have moved into turnkey projects precisely because the export of equipment by itself proved difficult. In products (e.g. simple items like small hand tools, various types of metal products, general purpose machine tools) and areas (mainly more developed countries) where the direct sale of products on their own has been easier, exports have grown without their 'packaging' in more complex forms of technology exports. In other words, turnkey exports provide an impetus to capital goods exports only for products and areas in which the sale of the goods on their own is difficult, slow-growing or restricted.

Let us conclude this section with a brief comparison of Indian turnkey exports with those of other developing countries. Evidence on other countries is, as may be expected, also rather patchy. Of the five other known Third World exporters of industrial turnkey projects – Argentina, Brazil and Mexico in Latin America, and South Korea and Taiwan in Asia[7] – we have only scattered indications on Brazil and Mexico. Both countries are setting up turnkey plants abroad in steel, paper, petrochemicals (and possibly cement and food processing), but nothing is known about the dimensions or values involved.[8] For *Argentina*, Katz and Ablin (1978) have collected data on 34 turnkey plant exports between 1973–7, but these do not exhaust total Argentinian activity in this sphere. Of the 34, only 27 can be counted as 'industrial' (defined broadly to include telecommunications). These are as follows:

TABLE 4.1 Argentinian turnkey plant exports, 1973–7 (US $ m)

Industry	*No. of contracts*	*Total value*	*Average value*
1. Food processing	15	44.9	3.0
2. Chemicals	5	45.8	9.2
3. Metal products	3	0.4	0.1
4. Oil pipeline and pumping	2	122.0	61.0
5. Communications	2	3.5	1.8
	27	216.6	8.0

SOURCE
Katz and Ablin (1978) p. 18.

Of the 27, 6 projects were carried out by affiliates of foreign companies, the rest by local ones. One relatively large project in oil pipelines ($120 million), won by Techint S. A. (an MNC affiliate), tends to distort the general picture. If we exclude this, the average value of the 26 contracts left falls to $3.7 million. In general, therefore, the foreign turnkey activity of Argentinian firms is fairly small scale, and the bulk of it is in fairly 'simple' technologies like food processing and metal products. Chemicals and communications represent a more sophisticated level of technical competence, but the most complex project (an automatic telephone exchange sold to Ecuador for $0.7 million) is quite small and is exported by an MNC affiliate. As in India, the exporters are mainly capital goods manufacturers which have moved into this activity to promote the exports of their products.

Information for South Korea is perhaps more complete, though still not fully comprehensive. Rhee and Westphal (1978) have collected data from newspapers on 15 turnkey exports up to the end of 1977,[9] of which 6 are completed and 9 are under construction, with a total value of $386.6 million. Of the 15, there are 2 paper plants, 4 chemical plants, 1 tyre plant, 1 cement plant, 1 textile mill, 1 rolling mill, 1 zinc smelter and 4 miscellaneous metal products. The average value of the completed projects is $2.4 million and of the ones under construction $37.4 million. The latter figure is pushed up by an exceptionally large ($235.1 million) cement plant being built in Saudi Arabia by Hyundai, a $60 million tyre factory in Sudan and a $72 million zinc smelter in Thailand.

None of the Korean turnkey exporters on record are MNC affiliates; nearly all of them are leading industrialists within South Korea – some 80 per cent of Korean turnkey exports are accounted for by the large trading/industrial houses. This is in contrast to Argentina, where a number of the local enterprises involved are of medium size. It appears that in many cases the turnkey exporters have acted as *organizers* rather than as direct providers of local industrial know-how. Where necessary, the process technology has been obtained abroad and Korean turnkey project exporters have, in such instances, contributed the civil construction portion of the work. As Rhee and Westphal remark, the large Korean exporters have been able to use their earlier civil construction experience (and contacts) to move into industrial project activity; where their own know-how has been deficient they have exploited their construction skills and supplemented it with foreign industrial technology.[10] It also appears that the larger houses have used their existing links with foreign companies with advanced technology to co-operate with them in overseas projects. Hyundai, for instance, has 40 such links in its domestic and foreign contracts.[11] The Korean case is thus instructive for two reasons: it shows that organizational skill can be exported for industrial projects even when domestic technological capability is lacking; and that skills can be transferred, given the existence of large, diversified firms, from one line of activity to another. Indian exporters, by contrast, have exploited a technological ability and have lagged in organization. None of their firms are structured like the Korean ones, so they may also have lost some of the learning 'internalized' by the latter.

Rhee and Westphal also provide some data on turnkey exports by Taiwan. Unfortunately, these only pertain to 1976, when 58 turnkey projects were exported, with a total value of $16.3 million and an average value of $0.3

million. These included 18 metal product plants, 11 paper plants, 8 plastic injection moulding plants, 7 water treatment plants and various others in activities like sugar refining, soap manufacture, batteries and air-pollution control. There are only two individual plants worth over $1 million – a sugar refining plant and a cement plant. The very small average size of Taiwanese plant exports places them in a different category from all the other Third World technology exporters and particularly from Korea, a country with which Taiwan is often paired. Its manufacturers appear to specialize in equipment of smaller scale (and perhaps of simpler technology[12]) than other semi-industrialized countries, though in terms of employment the exporting firms appear to be of reasonable size. As Rhee and Westphal (1978) note: 'Taiwan's turnkey plant exports through 1976 appear to reflect mastery of conventional production technology of intermediate capital intensity and for small scale production, and to be based on a comparative cost advantage rather than adaptations of technology absorbed from abroad' (pp. 10–11).

It is not only in terms of scale and sophistication that Taiwanese exports appear to be distinct. They are also heavily dependent on 'contacts with overseas Chinese' (Rhee and Westphal, 1978, p. 8) for marketing their technology in contrast to other exporters who have either won contracts in open international tenders (like India or Argentina) and/or have graduated up to industrial exports by establishing a market foothold through civil construction (like Korea). While the Taiwanese performance in international markets is often held to parallel closely that of South Korea, there are important differences between the two. In the field of industrial plant exports, in any case, the Koreans are involved in fewer but larger projects, and their expansion has been based on *sogo shosha* operations rather than on ethnic contacts abroad.

In sum, the evidence at hand strongly suggests that India is the leading exporter of turnkey industrial projects in the Third World in terms of the range, spread, size and sophistication of its exports, and in terms particularly of the *indigenously* assimilated technology embodied in its projects. It has successfully entered sectors, like large-scale power generation or electronic products where only developed country MNCs have till now set up plants (and it has sold generating plants to a developed country like New Zealand). And the diversity of activities covered is particularly impressive if viewed against its low levels of income and its rather lack-lustre export performance. A comparison with the recent progress of Korea, however, raises some questions about the organizational side of winning contracts abroad: a more aggressive policy, with a greater feasibility in buying 'missing components' of technology from abroad, may yield more dividends.

5 Direct Investments Abroad

The rise of 'Third World multinationals' has attracted some attention in recent literature.[1] According to the data at hand, a much broader range of developing countries is exporting this form of technology than any other. Almost every Latin American country, and a large number of Asian countries, are the homes of enterprises with some direct investment abroad (no doubt some African enterprises have joined in). While some of the data on developing country capital exports are difficult to construe as the export of local technology – tax havens like Panama, Bermuda and the Bahamas in Latin America and Hong Kong in Asia, seem to be large conduits for capital channelled by developed countries – it is clear that a considerable number of developing country enterprises have developed some source of competitive advantage that enables them to set up and successfully operate affiliates abroad.

The 'advantage' possessed by developing country enterprises may comprise one or more of the following: (a) the *technological* one, in the narrow sense, which would consist not, as in the case of MNCs from the advanced countries, of possessing new products or processes based on large R & D investments, but of having mastered mature techniques or having adapted them to the conditions of developing countries;[2] (b) the *managerial* one, of having the experience of operating in relatively primitive conditions and of having very cheap skilled personnel in managerial and technical cadres; (c) the *marketing* one of specializing in products, not of a highly promoted and differentiated variety, but of a kind especially suitable for Third World conditions (generally too 'simple' for the real MNCs) or (as in the case of Chinese firms from Hong Kong and Singapore) of an ethnic type to meet the need of particular immigrant communities.

The export of technology by means of direct investment may thus involve, at one extreme, the ability to reproduce a manufacturing technology learnt at home (and based on domestic capital goods), or, at the other, only the ability to organize and manage a plant based entirely on foreign technology and equipment, or some combination of domestic and foreign technical know-how. For countries with a large capital goods production base and local technical capability, foreign investments may involve a large element of indigenous technology; for those without such a base, such activity may rely mainly on managerial and marketing know-how. All direct investment, almost by definition, involve an injection of managerial skills; not all necessarily involve an injection of local technology.

Indian direct investments abroad have depended heavily on indigenous technology and equipment. Indeed, all of them have been in the form of joint ventures with local enterprises, with the Indian contribution being in the form of equipment made in India or capitalized know-how.[3] The government relaxed the policy somewhat in 1978 to permit some investment abroad in cash, but it is expected that these investments will continue to stimulate substantial exports of Indian machinery and intermediate products.

Table 5.1 presents some recent data on the values of Indian investments abroad and the earnings and associated exports (i.e. of intermediate

TABLE 5.1 Values of Indian foreign investment, earnings and related exports[a] (Rs. lakhs)

Period	*Equity exports*	*Dividends*	*Other repatriations*	*Related exports*[a]
Up to 1969/70	139.3	10.3	—	217.7
1970/71	39.7	2.3	0.9	30.3
1971/72	128.8	5.8	2.2	83.5
1972/73	86.3	13.4	7.2	123.8
1973/74	212.2	13.9	10.5	302.7
1974/75	266.6	22.3	14.9	710.1
1975/76	403.4	22.3	144.0	807.5
1976/77	315.7	35.3	93.5	949.9
1977/78[b]	171.4	—	27.3	705.6
Sub-total	1763.2	125.5	300.4	3935.1
From abandoned units	207.0	65.9	114.7	194.7
From units under implementation	65.9	—	—	125.0
Grand total	2036.1	191.4	415.1	4254.8

NOTES

1 lakh = 100,000.

[a] Related exports are of equipment, spares and intermediates *excluding* equipment provided initially for equity investment.

[b] Incomplete

SOURCE

Privately provided by the Indian Investment Centre.

products, spares, etc.) that it has given rise to. It shows that dividends comprised a relatively minor part of the foreign exchange contribution of overseas investments. Over twice the accumulated sum of dividends was

TABLE 5.2 Indian joint ventures abroad: in production or under implementation (1 Jan. 1979)

Activity	*S. and E. Asia*[a]	*West Asia*	*Africa*	*Europe*	*N. America*	*Total*
Oil, coconut processing	13	—	—	1	—	14
Food, drink	6	2	1	—	1	10
Rubber	1	1	—	—	1	3
Paper products	3	3	2	—	—	8
Paints, dyes	2	—	—	—	—	2
Pharmaceuticals	4	1	2	—	—	7
Other chemicals	8	2	—	—	1	11
Metal products	12	3	7	—	—	22
Wire and products	3	1	1	1	1	7
Non-electrical machinery and tools	6	—	4	1	—	11
Elect. products	2	2	—	—	—	4
Tractors, motor vehicles	1	—	1	—	—	2
Auto components	8	2	1	—	—	11
Clay, cement, glass	4	—	1	1	—	6
Wood products	—	1	—	—	—	1
Jute products	1	—	1	—	—	2
Textiles, yarn	12	—	8	1	—	20
Other	2	2	1	—	—	5
All manufacturing	88	20	30	5	4	146
Hotels, restaurants	5	1	1	6	5	19
Transport	—	2	1	—	—	3
Consultancy	4	7	3	—	1	15
Trading	—	4	3	1	—	8
Insurance	—	—	1	—	—	1
Construction	—	8	—	—	—	8
Total	97	42	39	12	10	180

NOTE
[a] Including Nepal, Bangladesh and Sri Lanka.
SOURCE
Indian Investment Centre, *Status of Indian Joint Ventures abroad and Guidelines*, (New Delhi, 1979).

provided by 'other repatriations' (royalties, technical fees and salaries), while the value of associated exports was several times higher. Associated exports were, in fact, more than twice the value of investments abroad, a fact which undoubtedly contributed to the government's decision to liberalize its policy.

Some details of the industrial and geographical composition of Indian foreign direct investment as of the beginning of 1979 are given in Table 5.2. The 180 ventures shown include 107 in operation and 73 under construction. A large number of applications are also presently under active consideration, and the relaxation of official controls is expected to lead to a rapid acceleration of the internationalization of Indian enterprises. The table also includes 37 non-manufacturing ventures: it may be noted that the number of foreign offices for certain service organizations, like consultancies or construction firms, is not directly related to the number of contracts won in these activities by Indian firms. We shall return to these activities below.

Of the 146 manufacturing ventures, the largest number (nearly 20 per cent) are in metal products (including wire products), followed by textiles and jute products (15 per cent), chemicals of all sorts (14 per cent), oil and coconut processing (10 per cent), automotive products and components (9 per cent), non-electrical machinery (8 per cent) and food and drink (7 per cent), with the remaining 17 per cent spread over various industries. These data are complete, since a record is kept of every foreign venture with the Indian Investment Centre.

These figures do not show a large number of joint venture efforts which have failed in the initial stages because of poor planning, over-optimism, poor performance or lack of sufficient knowledge of foreign conditions. Some failures are inevitable as part of the teething problems of operating abroad, but unfortunately the data with which they can be analyzed are not available. The government is aware of these failures, and has apparently undertaken measures to ensure better preparations for foreign ventures. A few comments on Indian capital exports are in order.

1. The table shows that Indian investments abroad span a broad range of technologies from the relatively simple (textiles, food and metal products) to the relatively complex (commercial vehicles, auto components, machinery, pharmaceuticals). The simpler activities predominate, as may be expected from the reasoning of the product cycle theory:[4] here the edge of Indian firms lies in mastery over the relatively standardized and low skill-intensive production technology, low operating costs and the ability to perform well under difficult developing country conditions.[5]

There is however, an increasing number of investments which, while they use relatively 'mature' technology, are in activities normally regarded as highly skill-intensive and marketing-intensive, and so beyond the range of Third World multinationals. A good example is TELCO's truck assembly affiliate in Malaysia. This firm assembles 7-ton trucks made entirely in India and sells them under the 'Tata' brand name. It sells more in Malaysia than Daimler-Benz which originally supplied TELCO the design, but which ended its technical agreement some 12 years ago;[6] it probably represents the first real automotive multinational to emerge from the Third World with its own trade name and technology[7] (there are, of course, several MNC affiliates using developing countries as bases for export). Its strength lies in the extremely rugged and reliable nature of its product which, despite its old design, is 'appropriate' for Third World conditions where some of the modern sophistication of driver comfort, speed, emission control, etc. are regarded as unnecessary by operators.

2. Most of the Indian multinationals are large producers on the Indian scene, though there are a number of medium-sized enterprises involved. Only two public sector enterprises have entered into joint ventures – Hindustan Machine Tools in Kenya and Balmer-Laurie in Dubai – but there are increasing pressures on them to move into this field from turnkey sales. As remarked in the previous chapter, some technology importers in the Middle East feel that the exporter has a stronger commitment to a project if it has an equity share in it. Thus, we find a reversal of the normal sequence of 'unpackaging' a multinational investment – the sale of equipment is gradually forced into a more 'packaged' form via turnkey projects to direct investment.

3. The promotion of sales of capital goods is a strong motive behind direct investment by equipment poducers. Among other manufacturers, however, the motives are partly to protect existing markets and partly to diversify away from limited or risky domestic markets.[8] Now that the Indian Government has realized the indirect foreign exchange earning capacities of overseas investments, there may also be various official financial and other inducements for spreading overseas. The export of capital goods and intermediates has, of course, always benefited from the normal export incentives offered by the government.

4. There does not seem to be any marked 'ethnic' complexion to Indian direct investments, either in terms of products or of destination. The products are standardized consumer and producer goods which could find markets in most developing countries. The main recipient of Indian investments is Malaysia (38 of the 180 in Table 5.1, or 21 per cent), where there is a fair-sized population of Indian origin; however, this population is not primarily engaged in business and there is little evidence that it has

provided any special benefit to Indian enterprises entering the country. In other countries of Asia, the ethnic connection is minimal; the other important host countries are Indonesia (18), Thailand (10) and Singapore (7), none with a significant Indian community. In the Middle East the United Arab Emirates have attracted 27 investments, Saudi Arabia 5 and Iran 4. In Africa the prosperous Indian business community may have attracted capital investment with Kenya and Mauritius (13 and 9 respectively), but with 11 going to Nigeria, this factor by itself could not have been very significant.
5. There are 22 Indian investments in the developed world. Of these 13 are in service activities (mainly hotels and restaurants) and 9 are in manufacturing. The manufacturing investments are as follows: Canada (rum, bicycle tyres and tubes); Netherlands (sal and mango oil, cocoa butter substitutes); Switzerland (garments); United Kingdom (cement products); United States (magnet wires; high density polyethelene products); West Germany (assembly of diesel engines); and Yugoslavia (steel wire ropes). This is an interesting collection of activities. Some (like cement products) are clearly to take advantage of locating near the market because of high transport costs; some (like garments) seem to make market entry easier in the presence of tariff protection; some (like sal and mango oil) are to exploit peculiarly indigenous technology; and some (diesel engines or wire products) may be to take advantage of the mastery of relatively simple and well-diffused technologies. It is rumoured that a number of Indian firms are exploring the possibility of investing in Eire to benefit from fiscal concessions and entry to the EEC market. The lack of a strong 'colonial' pull in manufacturing investment (i.e. to the UK) is worth noting.

In sum, then, Indian foreign manufacturing investments have taken place mainly to exploit its mastery over a range of production techniques coupled with a market cost advantage in technical and managerial labour. Its relative lack of a 'marketing' advantage is testified by the fact that over 70 per cent of its investments are in non-consumer goods activities where technical specifications and price competitiveness are likely to weigh more heavily than attractive presentation and advertising. However, Indian enterprises have made entries into such difficult consumer goods markets as pharmaceuticals and food products. And, as we noted earlier, some firms like TELCO have been remarkably successful in producer goods where faith in a brand name *is* of importance.

Success in foreign investing is a snowballing process. It gives the investors confidence to expand further and search for new markets;[9] it attracts other firms into going abroad; and it creates a better 'image' for Indian products, techniques and firms. The response to the recent liberalization of investment rules indicates that Indian firms will surge outwards strongly in the foreseeable future. More detailed research is, however, needed for us to assess how successful they have been and what their effects have been on the home and host economies.

Let us now compare Indian direct investment to that by other developing countries. In Latin America, the UNCTC (1978) records, for 1976, 203 parent companies in 22 countries; it is not stated how many are in manufacturing but presumably a fair number from the more industrialized countries are engaged in industrial activities.[10] The leading country is Argentina with 69 investments overseas, followed by Colombia and Venezuela (21 each), Mexico with 19, Peru with 16, Brazil with 15 and Chile with 12. The relatively low position of Brazil compared to Colombia and Venezuela is surprising, but may be caused by their different compositions of service versus manufacturing activity abroad.

In Asia (excluding Japan) Hong Kong leads the field in terms of overseas capital stock. Even discounting for funds channelled through the island by developed country firms, it appears that a number of local enterprises are active overseas in trading, banking and manufacturing activities. The manufacturing sector has spread abroad to take advantage of GSP privileges in the textile/garment industry (i.e. in order to export to Europe from countries like Malaysia, whose quotas are not fully used up), to protect existing markets for simple consumer goods (umbrellas, processed foods), to take advantage of lower labour costs and, occasionally, to procure raw materials (like timber).[11] One of the main advantages of Hong Kong firms has been the ability to use second-hand equipment (especially for textile manufacture) efficiently for small markets and unskilled labour. There is a very limited amount of local capital goods manufacturing in the island, but what there is seems to have contributed to the ability of its firms to produce more 'appropriate' technology. In the main, however, its competitive advantage lies in its operating experience and the low cost of its skilled personnel. This has also led other foreign firms, say from Japan, to launch joint ventures with Hong Kong firms in third countries.

Singapore is also a substantial investor in neighbouring countries. In Malaysia it is the second largest foreign investor and has spread across a number of industries.[12] Its firms tend to be much smaller and less capital-intensive than the MNCs from advanced countries, and they seem to cluster in activities with relatively simple techniques. The pressure to expand out of Singapore is stronger than for most other developing countries because of its relatively high per capita incomes, limited supply of labour and long experience of entrepôt trade.

Rhee and Westphal (1978) provide data on South Korean overseas investments as of end-1977. Of the total value of foreign investment of $71 million (in 229 ventures), manufacturing (17 ventures) accounted for 21 per cent of the capital, trading (126) for 20 per cent, forestry (7) for 20 per cent, construction (18) for 7 per cent, fishery (22) for 9 per cent and 'other' (39) for the remainder.[13] The manufacturing ventures are in simple consumer goods like textiles, garments, agricultural implements, food products and construction materials. About two-thirds of total Korean foreign investments went to other Asian countries, 19 per cent to North America (presumably mainly for trading activities), 10 per cent to Africa and 4 per cent to 'other' areas.

In general, the picture that emerges for Third World multinationals is as follows. The small fast growing island economies of Asia have led the field in terms of their 'internationalization': their limited domestic market, high export earnings, rising labour costs and threats to overseas markets have led their enterprises to invest broadly abroad. Other developing countries are less 'international' in this sense. Argentina is the major foreign investor in Latin America, and India in Asia, in terms of the number of operations overseas. A number of other countries have also given birth to a small number of multinationals.

In terms of the industrial and technological specialization of the Third World multinationals, most firms have gone abroad to exploit their knowledge of relatively simple and small-scale techniques and operations. In some cases – Hong Kong is perhaps the leading example – they have coupled this with quite highly-developed marketing abilities for consumer goods. The more diversified the indigenous engineering sectors of the home countries, however, the more complex the activities that its multinationals have entered. Thus, Korean firms seem to have a somewhat greater range than Hong Kong, Argentina considerably more than Korea, and India considerably more than Argentina.

While the lack of detailed information again prevents us from reaching a clear conclusion, it does appear that Indian overseas manufacturing activities span a wider and more complex range of techniques than those of other countries. They are relatively weak on consumer goods, and strong on intermediate and capital goods (further supporting the inferences which we have drawn earlier in this paper about India's technological development in the Third World). They are, mostly because of traditional Indian policy, based specifically on domestic know-how and equipment, as in Argentina but unlike the island economies. As the consciousness grows of their potential benefits to the Indian balance of payments, they are likely to grow in volume and size. The signs already exist that many Indian enterprises have developed the capability and the confidence to operate overseas successfully.

6 Licensing

We define 'licensing' very broadly here to cover three categories of technology exports: the sale abroad of patents, trademarks, blueprints and similar technology; the sale of 'know-how' not contained in such instruments; and the provision of training services.

Table A.5 presents the available information on the first two forms of licensing. It appears that, while the export of technology in this form is not as widespread as the others, it is not as small as was originally thought by the present author (Lall, 1979). I had argued earlier that the technological advantage possessed by developing country enterprises lay in the experience of the people who had mastered certain technologies in certain conditions, that this was not 'readily embodied in a saleable design, patent or blueprint'.[1] While this is partly true, it now appears that the licensing of designs, patents and 'know-how' by itself does take place to a significant extent from India.

The data in Table A.5 are far from complete, but they show that several patents have been sold abroad for processes developed in India. Some new pharmaceutical products have also been patented abroad and licensed to foreign producers,[2] but the details of this are not available and they have not been shown in the table. The licensing abroad of disembodied know-how has been more prominent, and has led to the assembly abroad of Indian trucks (TELCO once more) in several countries, of scooters (based on designs originally imported from Italy) in Taiwan, Indonesia and possibly Colombia, of bicycles and diesel engines. All of these have been based on the local assimilation of imported technology without necessarily requiring much R & D locally (though TELCO does invest in R & D). While this does not strictly count as a sale of know-how, it is interesting to note that recently a number of computer-programming firms have negotiated, or are negotiating, deals with European firms to undertake 'programme conversion' work in the West as well as to enter into joint ventures in other developing countries.[3]

The sale of patents, on the other hand, has been based upon local R & D efforts, as has the sale of process know-how. There are some striking examples of local innovation involving process know-how: reactive dye manufacture by Amar Dye-Chem, sold to a number of countries, including the US and Brazil; the adaptation by Bihar Alloy Steels of an imported smelter to do continuous casting of alloy steel, licensed back to the original German manufacturer,[4] the development of 'once-through' industrial boilers by Wanson, licensed to Vapor Canada Ltd, a subsidiary of Vapor Corporation, a US multinational. According to a recent magazine report, the publication of a list of firms able to provide technology has led to enquiries by a wide range of countries for several kinds of Indian technology.[5]

Many of these technological sales also involve the sale of equipment and/or intermediates from India. For the assembly of Indian products abroad the benefit to associated exports is obvious and is almost identical to the setting up of an affiliate abroad. For others, there may also be the stimulation of sale of crucial components. Wanson expects, for instance, to earn $2 million from its licensing arrangement with Vapor, including the provision of pressure parts and other components.[6] Some licensing may involve no further exports, of course, as with the sale of leather or chemical process technology.

The third form of technology exports under licensing (broadly defined) is the sale of training and similar services (technical, managerial, financial) abroad. As far as training is concerned, Indian enterprises are selling a variety of services: Tata operates one of two labour training centres in Singapore; the NIDC is setting up industrial estates in Guyana, equipping a Technical Training Institute in Malaysia, and has trained industrial consultants in Iran; HMT is establishing an advanced training centre in Iraq and an industrial estate to service a machine tool factory in Iran; Central Machine Tools (another government enterprise) has helped to set up a metal-working research institute in Iran; the National Research Development Corporation is setting up 15 pilot projects in Burma (for Rs. 25 m);[7] Star Textiles are setting up a textile training centre in Kenya.

As for the sale of technical and managerial services (excepting consulting, considered below) instances have been recorded of Indian firms winning management contracts for textiles, paper and cement enterprises. In a fairly recent contract HMT has accepted to provide several technicians to Alegeria. An unknown number of metallurgical specialists have been sent to Iran to help set up their steel industry. Several enterprises provide training facilities for foreign technicians in India.

There is very little information available on licensing and similar services abroad by other developing countries, though there is little doubt that such services are rendered. In Mexico, for instance, Katz and Ablin (1978) mention that certain processes developed locally – for direct reduction of steel,[8] for a petrochemical process, for construction materials and for newsprint manufacture – are being used in several countries, ranging from Latin America to Africa and Asia.[9] In part, these technologies have been sold in the form of turnkey projects, but it is likely that some transfers have been made through licensing. The same authors suggest, however, that licensing of patents has not been widely practised by Argentinian firms, at least for their sample of the major technology exporters (who exported their technology by direct investment or by turnkey projects).[10] This does not exclude the possibility that technical services, more broadly conceived, are being rendered to foreign countries by Argentine firms – given their relatively advanced know-how and experience it would be surprising if this were not so.

Rhee and Westphal (1978) note that 'only isolated anecdotal information' is available on Korean activities in this category. They record one case of the licensing of manufacturing technology, for nylon tyre-cord production to Thailand and Taiwan; they find several instances of exports of technical services for agriculture (irrigation and rural programme management). A recent newspaper report[11] says that the Hyundai group, one of the leading integrated enterprises, is negotiating with India to help in the construction of a modern shipyard (for repair), and to provide follow-up training services. Hyundai is already a world-class manufacturer and exporter of ships from South Korea, and reputedly possesses some of the most modern shipyards. The same report notes that South Korea (the firm involved is not mentioned) is also examining a proposal for setting up an iron-ore pelletization plant in India.

Nothing is known about licensing and similar activities by Brazil, Hong Kong, Singapore or Taiwan. What can we conclude on the basis of the existing evidence? India is clearly one of the leading exporters of technology in this form; precisely how it ranks in relation to other developing countries is impossible to say. The fact that some of its innovations (minor though they may be) have been licensed for use in developed countries says a great deal about the underlying technological capabilities of its enterprises. Much remains, however, to be learnt about the provision of technical assistance by India and by other countries.

7 Consultancy

In itself, the sale of engineering consultancy services abroad is an almost purely disembodied form of technology export. The nature of the activity is such, however, that it is almost bound to lead to the exports of associated equipment from the home country, since industrial consultants usually form a direct link between the investor and the equipment supplier.[1] We shall not, for lack of data, go into this aspect of engineering consultancy, but confine ourselves to the evidence on consultancy sales as such.

Engineering consultants can be of several types. They can be extremely specialized, by industry or by process, or they can be general engineering consultants, capable of dealing with a wide variety of technologies. Given firms can perform a very narrow range of services in a given situation, or they can undertake the entire design, construction supervision and commissioning of a project. They can operate at a relatively 'low' level of technical sophistication, by conducting project supervision or detailed engineering; or they can reach 'high' technological levels by conducting basic design work for complex processes. The main determinant of the technological range and level achieved by a country's engineering consultants is the relevant experience they have gathered at home. As we shall discuss later, the broader the range of investment activity based on indigenous technology and the further back along the vertical chain a country has gone in terms of building the capital goods required for investment, the greater will be the capabilities of its engineering firms.[2]

India has over 100 professional consulting firms; some 50 consulting and design engineering organizations and about the same number of other more specialized consultants. It seems that a large number of them are active abroad, but we have managed to collect information (and even this is incomplete) on technology exports by only 20 of them. This is displayed in Table A.6.

The first two enterprises in the table, NIDC and EIL, large public sector companies, are general engineering consultants with an enormous spread of overseas activity over region and industry. NIDC, in particular, has undertaken everything from town-planning in Tanzania, general economic planning in Libya, minor technical services in Italy and the UK, and mineral survey in Afghanistan to various types of industrial work in several developing countries. EIL is more specialized (but not exclusively so) in chemical and petrochemical industries, where it has undertaken independent work in the Middle East as well as sub-contracts from developed country enterprises.

Among the more specialized public sector firms the most remarkable are WAPCOS and RITES. WAPCOS now conducts 85 per cent of its total business abroad, in the design, construction and supervision of hydro- and irrigation systems and related consultancy services. In 1977/8 its foreign earnings came to $2.4 million; these earnings have been growing at about 40 per cent per annum in recent years. RITES gets 30 per cent of its earnings from foreign consultancy work in railway construction and management, about $280 thousand in 1976/7.[3] It has performed services in a variety of Asian and African countries, and has even executed a contract in New Zealand. Its biggest award has been a $25 million contract for managing the entire railway system of Nigeria. It is presently negotiating, jointly with two firms from developed countries, a major construction contract in Jordan, having lost a contract in Iraq to the Brazilian firm Mendes Junior. It has also entered into an agreement with the Algerian government to render a wide range of services, studies, personnel and construction work.

The experience of MECON, a large public sector metallurgical consultant company (with some 2000 professional engineers) is also worth noting. MECON gained its experience basically in the construction of Soviet-aided steel mills in India: there was so much 'trouble shooting' work involved in adapting and improving these technologies that Indian engineers quickly accumulated a sizeable fund of valuable know-how.[4] MECON recently entered the international market, and has conducted feasibility studies for a number of countries, including Hungary. Its biggest foreign contract is in Nigeria, where it is in charge of the technical consultancy and project monitoring of a large steel mill with the constructors being a West German firm.[5] More interestingly, it is entering a collaboration agreement with Alusuisse, a large metallurgical multinational from Switzerland, to bid jointly on contracts in Europe as well as elsewhere. The Swiss firm will apparently specialize in the aluminium, and MECON in the iron and steel, side of their operations: a testimony to the international competence and competitiveness of the latter. It is expected that this agreement will enable MECON to strengthen its aluminium know-how and to enter European markets where the lack of an established name can be a crippling barrier.

A number of public sector manufacturers are also actively selling consultancy services abroad. HMT has sold various types of services in countries ranging from Algeria to the Philippines. IDPL is starting to provide services on pharmaceutical manufacture, mainly to the Arab countries. The Fertilizer Corporation has conducted marketing studies in Burma (an $800,000 contract awarded by the World Bank) and the Philippines. Other enterprises, like HUDCO (Housing and Urban Development Corporation) and KVIC (Khadi and Village Industries Commission) are entering foreign markets to sell their specialized services, but they have not been shown in the table for lack of specific information. The Oil and Natural Gas Commission (ONGC) is actively engaged in oil exploration for the governments of Iraq and Tanzania.

Information is even more patchy as far as private sector consultants are concerned, but some of the larger enterprises are mentioned in the Appendix table. Tata Consulting Engineers, a general engineering firm but with a strong interest in the electrical power sector, gets 40 per cent of its revenue from abroad. Its overseas earnings in 1976/7 came to just over $2 million (when it won the 'Outstanding Export Performance in Consultancy' award), and were growing at around 200 per cent per annum. Of its many assignments abroad it has executed one in West Germany to provide design services for a thermal power station to Deutsche Babcock-Wilcox.

Dasturco is the largest private sector consultant for iron and steel. It is involved abroad much more extensively than MECON, its public sector counterpart, with experience in 26 foreign countries, including West Germany. Its initial entry into foreign markets was facilitated by contracts awarded by UNIDO, which enabled it to build a reputation overseas.[6] It now has permanent offices in Düsseldorf and New York, and has also worked in Latin America. The two major areas of its current involvement are Libya and Venezuela.

A number of other private firms are also involved overseas, and by all indications consultancy activity is growing fairly rapidly. It appears that a number of Latin American engineering consultants also operate abroad,[7] but no details are available on their activities. Mexico has a well-known general consultant, Bufète Industrial, which provides services overseas and has also developed a manufacturing process for newsprint which is being used abroad.[8] There are also petrochemical and metallurgical consultants operating from Mexico, and Nadal (1977) remarks on their entry as subcontractors for detailed engineering work to developed country consultants. For Argentina, only indirect evidence, from the discussion of Katz and Ablin (1978, pp. 37–8), exists that engineering consultants are promoting the sale of Argentinian technology abroad. For Brazil there are occasional signs (like the Mendes Junior contract in Iraq) that its firms are active far afield; it is very likely that they are more powerfully in evidence in neighbouring Latin American countries.

For other Asian countries there is no direct evidence of any industrial consultancy work internationally for any country besides India. It is, however, likely that some consultancy exports do take place, especially from South Korea where the large integrated companies may have spawned consulting offshoots.

As with other forms of industrial technology, only a tentative conclusion can be drawn from the limited evidence at hand; it appears again that Indian firms lead the Third World in international sales of consultancy services. In fact, a number of these firms have achieved such a degree of internationalization that it is likely that they are able to tap and provide technologies not in use in India (this seems to be the case with iron and steel technology where Indian consultants sometimes advise the procurement of foreign equipment). In the main, however, they probably tend to act as a medium for the transfer of Indian technology and Indian equipment, and their experience overseas no doubt feeds back into their work within India.

8 Non-Industrial Technology

The sort of technology embodied in construction and service exports is of a totally different nature from that required for manufacturing industry. It relies much more heavily on organizational and managerial expertise than on an understanding of scientific, metallurgical or mechanical principles, though construction work obviously requires some very advanced forms of engineering knowledge. The processes of 'technological' development and learning are, therefore, of a completely different nature for non-industrial as compared to industrial activity. This does not render them any less relevant for a study of development issues: it simply calls for a different sort of analysis and different sets of policy recommendations.

This study has concerned itself primarily with industrial technology. The section on non-industrial technology exports is correspondingly shorter, because of its lesser relevance and because of the scarcity of data. The next two sub-sections deal with civil construction exports and the exports of other services from India.

I CIVIL CONSTRUCTION EXPORTS

A large number of public and private sector firms are active in construction activity overseas, particularly in the Middle East. The public sector firms include EIL (Engineers India Limited) which, as noted above, is also active in industrial plant construction; this firm has projects in hand for Rs. 7000 million (about $850 m) in total, of which $570 million are overseas.[1] The largest enterprise in civil construction exports is Engineering Projects India (EPI). EPI sales have been growing at over 200 per cent per annum, and it is operating, not only in the Middle East but also in Thailand, Yugoslavia and in various parts of Africa. Other public sector agencies include the National Building Construction Corporation, Hindustan Steel Construction Limited, the Indian Road Construction Corporation, the Housing and Urban Development Corporation, and the International Airports Authority of India.[2] The private sector firms are too numerous to name, but are increasingly active in the construction of hotels, hospitals, roads, palaces and drainage systems.

One of the major customers for Indian construction services has been Libya where airports, roads, townships, hospitals, sewage works and the like are being built throughout the country. Other countries include Iraq, where Indian companies are building the Najaf-Kufa sewage system, the Baghdad University, roads and bridges; the Gulf States, where they are building a palace complex, an airport, complete townships, military camps and drainage systems; Saudi Arabia, where hotels and industrial complexes are being constructed.

To quote a recent report: 'In all, Indian companies have won an impressive $1.2 billion in construction contracts in the Middle East in the past three years. And there is considerable optimism for the future in view of the sharply rising business curve which has shown contracts rising from $91.9 million before 1977 to $384.4 million in 1977, then to $741.6 million in 1978.'[3]

India does, however, lag well behind South Korea in this sphere of technology exports. In the Middle East, which accounts for nearly 80 per cent of its overseas construction activity, Korean contracts by the end of 1977 exceeded $6.7 billion, and those awarded during 1978 reached $9 billion. Its receipts came to $1146 million in 1977, and were projected to rise to $2079 million in 1978 and $2792 million in 1979.[4] Other developing countries, like Taiwan, Pakistan, Philippines and probably Egypt, are also active in the Middle East, but their shares are likely to be much smaller than that of India. In Latin America numerous construction firms are active in neighbouring foreign countries, but with some Brazilian firms after an exceptionally determined drive, having reached Africa and the Middle East.[5]

The export of construction technology requires three sorts of expertise: of having undertaken large projects at home; of having sold, organized and implemented projects overseas; and of putting together an attractive financial package (with credit, guarantees and the like). Korean firms have been fortunate in all these respects. As Rhee and Westphal note: 'aspects of modern construction technology were learned by Korean firms through their involvement in US military construction projects in Korea and in various South-east Asian countries (most importantly Vietnam), which predates their venture into the Middle East. . . . Moreover, 'marketing' is done by the Korean firms, acting without foreign agents. This is one area where Korean know-how relating to transactions is second to none'.[6] It will remembered from chapter 14 that Korea has 18 construction affiliates abroad;[7] it has also set up the Korean Overseas Construction Corporation to coordinate the activities of its construction companies.

Indian firms have had considerable experience of large-scale civil works within their country, though perhaps without the benefit of an injection of the latest American know-how. They have, however, been deficient with respect to the 'marketing' of their product overseas; it is the recognition of their lack of aggressive selling that has recently led the Government to encourage the setting up of consortia to bid for particular contracts and of regional offices in Saudi Arabia, the Gulf and elsewhere. As the *Financial Times* report quoted above says: 'Business has been less forthcoming in Saudi Arabia, with Indian concerns discovering that major projects there are seldom advertised. In the hope of generating some business, the Association of Indian Engineering Industry has opened an office in Saudi Arabia to gather marketing intelligence and made bids on behalf of its members.'

The lack of proper financial facilities has also contributed to Indian construction exports lagging behind Korea's, whose larger firms receive the wholehearted financial (and political) backing of their government. As a *Financial Times* report noted last year: 'Well over 30 South Korean contractors operate abroad, in sharp contrast to the position a decade ago. Their efforts have been greatly assisted by a Government which has put its full financial weight behind the export drive.'[8] And again, in its survey of Saudi Arabia: 'In finance for bonding and operations, Far-Eastern companies have a slight edge because of the close relations they enjoy with their governments. The Saudi banks, too... have also been prodded by the Government into providing contract finance for South Korea and Taiwan.'[9] For Indian contractors, on the other hand, 'many more contracts would have been won but for the limited experience in handling such overseas contracts and inadequate financial support.... Sometimes, even if the companies want to take risks, their bankers do not approve. The restriction is such that if bankers issue a veto, no contract can be taken up by Indian construction companies.... The problems faced by the construction companies are many: tight exchange control right from the stage of opening an overseas account, difficulties in importing machinery from third countries, and cumbersome procedures for sending labour to overseas jobs.'[10]

While bureaucratic handicaps are probably less now with the Indian government adopting a more aggressive promotion policy, it is unlikely that Indian firms receive anything like the all-round backing given by the Korean (and Taiwanese) governments. Another important factor behind Korean success has been the disciplined and high quality of its construction labour. As the *Financial Times* noted: 'Eighty percent of the 30,000 to 40,000 Koreans in Saudi Arabia have done military service – they are recruited by the companies when they are discharged. Their industry and conscientiousness – and the pride they take in their project – are a continuous source of wonder to Saudis and foreigners.'[11]

The very efficiency of its labour force may, however, be helping Korea's competitors; the rapidly rising wage costs of labour has made the Koreans 'well aware that the best times have indeed passed for the Middle East construction industry. . . . There is no doubt, also, that Koreans will be looking increasingly to non-Middle East markets. At present 10 per cent of the industry's overseas business is in the Far East, 4–5 per cent in Africa and 7–8 per cent in South America.'[12] Thus, countries with a labour cost advantage, in particular India, should expect to take a bigger slice of the cake. Korean firms, given their diversified structure, should be able to retain some of their advantage by moving into more complex industrial activities.

The recent rapid expansion of Indian overseas construction contracts shows that they are 'learning' how to use their advantage and to overcome some of the various barriers of red tape and financial stringency. However, there is some way to go before the leading contractors are given sufficient financial muscle to reach Korean levels of performance in this sector. It is also unlikely that they will be permitted to reach the high concentration levels (top 10 firms account for half of Korean turnover) found in Korea, or to diversify to the same extent as the Korean trading manufacturing houses and so to benefit from the various oligopolistic advantages that size confers.

II OTHER SERVICE EXPORTS

There are two other forms of overseas activity by Indian enterprises which may be mentioned in the present context though they do not amount to very much in terms of their magnitude: overseas banking and hotel operations. Most of the larger Indian banks have opened branches overseas,[13] mainly in order to service Indian communities abroad, in Asia, Africa, Europe and N. America. The leading Indian hotel chain, Oberoi, which originally had collaboration agreements with foreign hotel chains, now has over a dozen branches overseas in Asia, Africa and Europe. Other hoteliers are becoming active overseas, and have even set up in Eastern Europe.

We have already mentioned the construction overseas of hospitals (a form of 'health technology' export)[14] and airports, roads and railways (in the transport sector). Some Indian firms have undertaken agriculture and irrigation projects overseas. Voltas, for instance, has got 20 projects in hand for water well drilling, drip and sprinkler irrigation systems and power distribution in countries ranging from Iraq and Mauritius to Indonesia and Malaysia. The overseas work of WAPCOS (in large-scale water and power engineering) has already been noted, as has that of the Khadi and Village Industries Commission's export of village-industry technology, and the

National Small Scale Industries Corporations' selling up of 32 small-scale industries in Tanzania. Other government organizations are exporting particular types of agriculture-related technology. Outside the market system, of course, there are numerous official schemes for technical assistance to other developing countries in all of these sectors.

Other developing countries are also exporting various types of service technology. Hong Kong is a major international banking centre and its leading banks are highly multinational in their own right. Most other developing countries of any size have extended their banking systems overseas, as they have their hotel networks. Agricultural services are also being exported by many countries. Brazil and Mexico are selling petroleum exploration technology. Trading companies from developing countries, especially from Korea and Hong Kong, are spreading actively over the globe.

India thus shares, but in a relatively modest way, in the internationalization of all these various activities. The only one in which it may have a clear lead is in small-scale and village industries where it has accumulated a fund of experience and conducted a great deal of research. For others it is impossible to pass a judgment without examining much more evidence on Indian and other developing country activity.

This concludes our survey of the exports of non-industrial technology. The remainder of the book will focus upon manufacturing technology and will touch only peripherally upon the exports of other technologies. It will seek to explain the nature of the technology that is exported by Indian enterprises, the determinants of the technological development that gives rise to exports, the limitations on developing country technologies, and the costs and benefits to India of technology exports.

Part C
Assessment

9 Nature of Exported Technology

I INTRODUCTION

An assessment of the growth and implications of technological development in, and technology exports by, developing countries should ideally be undertaken after a comprehensive collection of data and a detailed examination of the individual enterprises involved in the technology transfer process. The existing literature on this subject is in its infancy. The vast literature on the history of technical progress in developed countries,[1] on the 'pure' theory of technical change and recent 'evolutionary' approaches to innovation,[2] on the relationship between growth, productivity change and innovation,[3] and on the micro and macro-economic conditions for innovation,[4] offers some extremely valuable insights. The focus of these enquiries has, however, been rather different, and, in the absence of empirical testing, it is difficult to judge how relevant their findings are to the conditions of the developing countries.[5] General theories of 'latecomers' catching up with the advanced nations along Gerschenkronian lines,[6] while appealing in the broad sweep of their historical inevitability, leave unanswered many questions of detail about how different enterprises, industries and countries 'catch up', where they fail, and what to do about it.

The present paper can only advance some tentative hypotheses based on rather scattered information of the sort described previously. Some of these may be vindicated by further analyses and research, others may be modified or rejected. They are presented in this exploratory spirit.

This chapter starts (Section II) with a brief analysis on the 'innovative' activity which gives rise to technical advance on the part of developing country enterprises, drawing on recent theories of 'evolutionary' technical progress to analyse their technical activity. It then goes on, in Section III, to distinguish the characteristics of technological advance in developing countries which give rise to exports, as compared to those which do not. It closes, in Section IV, with a classification of the different sorts of technologies exported, and their implications for the importing countries.

The more general determinants of, and limitations to, technical progress in developing countries are discussed in the following chapter.

II 'INNOVATION' IN DEVELOPING COUNTRIES

There are two factors which have worked against a proper appreciation of the nature of innovation in developing countries. One is the dominating concern in the relevant empirical literature with major breakthroughs rather than with minor changes as the source of technical progress. As Rosenberg (1976) and others have noted, this 'Schumpeterian syndrome' has tended to detract from the important contribution to technical progress made in the diffusion, adaptation and application of new technologies. The other is the theoretical characterization of technical progress in neo-classical economics, with a given production function depicting an array of efficient techniques along which an enterprise can shift easily in response to changes in factor without registering any technical progress, and shifts which (caused by forces which are never clearly specified) lead to the utilization of more efficient processes that constitute technical progress. The limitations of this approach as the portrayal of reality have been extensively analysed.[7]

In part the exclusion of product innovation, which contributes as much (or more) to modern technical advance as process innovation, is a crippling restriction.[8] In part, however, even the analysis of process innovations is seriously deficient. Enterprises do not confront an array of different techniques between which they can switch easily: they operate, for reasons of historical accident, with particular technologies which are difficult to change, and which they cannot change without costly effort to acquire and implement new knowledge. It is misleading to distinguish between movements along a production function from shifts of the function itself. Techniques and products evolve with experience, modifications and investments in knowledge creation: according to Nelson and Winters' evolutionary theory, 'almost any non-trivial change in product and process, if *there has been no prior experience*, is an innovation'.[9] We can differentiate between minor and major innovations, of course, on obvious intuitive grounds, but they are simply different points along a continuous spectrum of technical activity, which includes imitation as an innovative activity at one end and fundamental research at the other.

Innovation in this sense obviously does occur in developing economies. Techniques are adapted and improved after they have been imported; new techniques and products are developed from old ones; and sometimes even major innovations are made in terms of breakthroughs based on original research work. On the whole, however, innovation in developing countries is 'minor', based upon basic scientific and technical advances made abroad, in the developed countries.

Perhaps even this concept of technical progress is too narrow if we consider the whole set of activities involved in *implementing* a given technology in an operational plant. The narrow definition of technical progress applies to individual products or processes. The implementation of an entire technology, e.g. a turnkey plant, or a direct investment, involves a large set of products and processes on the manufacturing side, and an equally large set of activities on the organizational and financial side. There is no reason why the concept of innovational activity – with 'learning' based upon experience and deliberate investments in knowledge creation – cannot be applied to these non-technical activities.

The determinants of innovative activity at the firm level have been analysed theoretically by Binswanger (1974) and empirically by a range of different economists with different perspectives and tools.[10] The theoretical work generally assumes a profit maximizing entity operating within an 'innovation possibility set associated with a given level of expenditure or with different elements associated with costs of different amounts. Firms are assumed to choose the profit maximising element.'[11] As the proponents of the evolutionary theory note, this neo-classical approach has analytical attractions, but it does not come to grips with one of the major elements inherent to all innovative processes.

This element is *uncertainty*.[12] Basing their observation on empirical findings, Nelson and Winter argue, with considerable persuasiveness, that

> . . . everyone agrees that R & D is an uncertain business. Uncertainty resides at the level of the individual project, where the 'best' way to proceed seldom is apparent and the individuals involved instead have to be satisfied with finding a promising way. Uncertainty also resides at the level of R & D project selection. The enormousness of the set of possible projects, the inability to make quick, cheap, reliable estimates of benefits and costs, and the lack of convenient topological properties to permit sequential search to home in rapidly on good projects independently of where the search starts, means that project choice, as well as outcome given choice, must be treated as stochastic. The problem with the maximisation metaphor is not that it connotes purpose and intelligence, but that it also connotes sharp and objective definition of the range of alternatives confronted and knowledge about their properties. Hence it suggests an unrealistic degree of inevitability and correctness in the choices made, represses the fact that interpersonal and interorganizational

> differences in judgement and perception matter a lot, and that it is not all clear ex ante, except perhaps to God, what is the right thing to do.[13]

Nelson and Winter propose, on the basis of these considerations, that R & D projects, be viewed as 'interacting heuristic search processes', where a 'quasi-stable commitment to a particular set of heuristics regarding R & D can be regarded as an R & D strategy'.[14] Thus innovation (extending what Nelson and Winter say about R & D to all technological activity) is essentially an *uncertain search* activity. There are certain points of reference, certainly – a knowledge of demand conditions in the market on one side, and of scientific potential on the other, though both the 'demand-pull' and 'supply push' influences on innovation are subject to their own uncertainties. These points can lead to innovative effort along certain natural 'trajectories', or, as Rosenberg terms it, 'technological imperatives'. These trajectories or imperatives are based, not on innovation possiblity frontiers in a sense that implies certainty, but 'technicians' beliefs about what is feasible or at least worth attempting'.[15]

Nelson and Winter go on to discuss two natural trajectories which have been exceptionally significant and widespread – the exploitation of latent scale economies and the increasing mechanization of operations done by hand. They also discuss at length the significance of the *institutional setting* in which innovations are made and exploited, pointing out that institutional differences between industries lead to different rates of technological progress and diffusion. The existence of profitable opportunity, market demand, official policy, costs of imitating other firms' innovations,[16] all enter the picture to influence the selection of potential innovations for practical application.

We have gone into these rather abstract issues at some length because they contribute to our understanding of 'minor' innovation in developing countries. Given the uncertainty inherent in any purposeful technological activity what are likely to be the distinctive features of innovation in developing countries?

Most developing country enterprises, especially locally-owned ones, are not on the frontiers of technical advance as set by the industry leaders in the developed countries, but some distance behind it. The distance itself differs from industry to industry, depending upon the developing country's length of experience with the activity, its investments in absorbing new technology and the rate of change of the technology in advanced countries. This means that, in general, the bulk of technological activity in developing countries – from small changes on the shop floor, minor modifications of products or processes to full-blown design and development work[17] – will be devoted mainly to searching out, adapting and implementing those aspects of technical advance abroad which are likely to be profitable in local conditions. It will, in other words, be technological activity essentially related to the diffusion and imitation of foreign technologies, but with certain important differences from the conditions under which such diffusion and imitation occur in advanced countries: relative factor prices are very different; specific factor availabilities for technical work are different; market and operational conditions are different; raw materials are often also different; and supplier capabilities are much more limited. Let us take these in turn.

First, relative factor prices in the usual sense of physical capital *vis à vis* labour, obviously favour innovation in developing countries towards the use of more labour intensive techniques. While the existing evidence on such technical advance is mixed,[18] it is undeniable that a number of 'peripheral' processes (storage, transportation) are run much more labour extensively in developing countries, and that, in the application of new 'core' technologies, very high degrees of automation are screened out. Furthermore, adaptation often takes place in a passive manner: enterprises run old techniques (initially imported in an unadapted form) for much longer, simply incorporating only the new elements which are rational in the given economic conditions.

Second, the agents which undertake technological work are likely to be available in different quantities as between developed and developing countries. For most developing countries, there is likely to be a relative shortage of trained technicians and engineers, especially those capable of undertaking basic design and development work. This will lead to a slower rate of minor innovation, in given conditions, than in developed countries. For some developing countries (in particular India), however, there may be an *oversupply* of trained technical personnel at all levels of manufacturing activity. This is likely to lead, once a threshold of experience has been reached, to a more rapid and effective translation of production experience into useful innovation.[19]

Third, the different demand conditions in developing countries are likely to bias product innovation in certain direction: in consumer goods, to simpler design, less sophistication and perhaps longer life; in capital goods, to less specialized equipment, greater ruggedness and easier maintenance and operation. And, in both, scales of operation are likely to be smaller. Again, this will operate partly by screening out innovations abroad which are likely to prove too costly in developing countries; in part, however, it will be the result of active design and development work. Both are search processes involving effort and expense; even the selection of 'appropriate' innovations from a shelf of changes made abroad requires considerable technical research and development work.[20] This is true especially of mechanical engineering industries, where patents by the innovators are not very helpful to an imitator, and where a great deal of experimentation is required to achieve the right specifications, tolerances, combinations, etc. for a new product. It may be noted that innovations in machinery will be made mainly by the capital goods sector, but they may occasionally also be made by the users, in two ways. First, minor modifications can be made by users without much special equipment or investment; second, many industries in developing countries with strict import control on machinery (India until recently) set up their own equipment building shops. Thus, a food processing multinational in India makes a lot of its own equipment, and has reached such levels that it is providing assistance to Brazil within the company frame work.[21]

Fourthly, the need to use local raw materials, which may be different from those abroad, is often one of the main spurs to innovation in developing countries. A greater part of R & D work in food processing and chemical industries in India goes into the effect to import-substitute raw materials,[22] and the lack of certain kinds of material leads to changes in product design.[23]

Finally, the lack of a fully developed, specialized component supplying industry leads to a great deal of technical work for the user industries, to develop substitutes, to build up supplier capabilities, to simplify specifications, and generally to take a much more active role in component design/production technology than in developed countries.[24] In some cases, the pressure to reduce dependence on single suppliers (who may have imported advanced technology themselves) leads enterprises to 'innovate' hand in hand with new suppliers so as to bring their products up to specification.

There are, therefore, major institutional and environmental differences, even within 'minor' innovative activity, in developing, as opposed to developed, countries. In some ways, innovation involves less work in developing countries because they are 'merely' catching up with what has been done abroad. However, in many others, it involves far more work. The lack of experience and infrastructure in developing countries renders local technical effort a much more risky and path-breaking task. The different demand–supply conditions, prices and skills are likely to force innovators to do more than 'merely' copy.

It is likely, in this context, that these different conditions allow developing countries to innovate more on improvements of *process* technology than on *product* technology. In the machinery sectors the distinction may not be so important, but in others (e.g. vehicles, pharmaceuticals) the two require different types of market conditions and R & D investments. Thus, to produce a completely new product (e.g. a new engine or a new drug) may be more difficult and expensive than finding a more efficient way of producing a given engine or drug (especially if local raw materials are different from those for which the original technology was designed). Furthermore, it may require a larger market than most developing countries provide and more far-flung marketing networks than most enterprises there possess (most major product innovations in developed countries are made by MNCs *for* world markets). Much depends, of course, on the nature of the industry and the technology in question: some products evolve slowly, other in quantum jumps, and for the former, developing countries may well be able to produce effective innovations (even if this involves simplification and less performance).

For all these reasons, therefore, developing country innovation will have, in Nelson and Winter's terms, a different set of 'natural trajectories' and these involve as much uncertainty and search as innovation elsewhere. We shall return to some broader determinants of technical progress in developing countries in the next chapter.

III TECHNICAL PROGRESS AND TECHNOLOGY EXPORTS

Given the different 'trajectories' of technical advance in developing countries, not all industries in a given country are likely to experience equal rates of progress, not all enterprises within a given industry are likely to advance, and not all kinds of technical advances are likely to lead to technology sales overseas. The fact that different industries enjoy different rates of productivity increase is one of the best known facts of economic history, and has exercised many analysts and econometricians to look for the causes for the differences. We have nothing to add to this particular debate here, and leave it to future research to decipher the extent to which the causes of productivity advance differ between rich and poor countries.

We may, however, advance some hypotheses about the relationship between different forms and stages of technical progress and the nature of technology exports. Let us regard all progress in technology and efficiency as a form of 'learning'. The use of the term 'learning' for the process of advance is intended to emphasize two aspects of the process: it necessarily takes time, and it involves experience of right as well as wrong steps. It is not intended to convey the impression that it is a costless process. On the contrary, learning involves all the expenses involved with 'uncertain search'. Besides its classic associations with a certain sort of technical progress, the term 'learning' catches much of the flavour of the process involved in developing countries.

We may distinguish between two broad types of learning that contribute to productivity advance: technical learning proper, and learning of non-technical (managerial, marketing, financial, etc.) aspects of production. Let us consider them separately.

a. Technical Learning

For any given technology (or family of technologies) we may distinguish between three stages of technical learning, each with two sub-stages:

i. *Learning within a given technology*

When a new technology is imported, two sorts of learning can contribute to greater efficiency in its use over time, neither of them requiring major modifications in its 'embodied' technology, i.e. the equipment and machinery utilized. These are: simple 'learning by doing', whereby workers become more efficient simply through experience; and 'learning by adaptation', whereby small changes, made by shop-floor technicians, engineers and managers, to the product or process contribute to productivity. The best documented and researched study of the significance of such learning is on Argentina (Katz, various); the evidence suggests strongly that such activities as 'troubleshooting', rearrangement of plant, adaptation of equipment, and so on – none of which fall within the strict definition of 'technological effort', normally taken to be formal R & D – contribute greatly, and continously, to increasing efficiency and productivity.

ii. *Learning the embodied technology*

The next stage of learning occurs when some of the machinery itself is manufactured within the country, enabling it to reproduce or improve that particular function within the technology. Here we may distinguish between 'learning by imitation', where local engineers simply replicate foreign designs and blueprints, and 'learning by design', where they progress to understand the basic scientific and engineering principles involved and so are able to adapt, change and improve the machinery. There is a small but important step between these two stages of learning: the first may consist of simply putting together imported components, or doing some detailed engineering, without enabling local manufacturers to understand and reproduce the equipment. The second necessarily involves basic design work, a knowledge of the exact qualities and specifications of the materials and components used, and of the mechanical principles involved, and so requires considerably more experience, search and purposeful technical work. It usually requires the setting up of a separate R & D department.[25]

iii. *Learning the entire technology or production system*

The final stage of learning a technology is to graduate from the capability to produce particular machines to one to reproduce the whole technology in a functional plant, with a variety of different kinds of machinery, each with its own sub-technology, with an overall design suited to particular needs, and with the ability to train others in the know-how needed to run the plant. This capability can only result from an extensive understanding of a technology as a whole which comes from considerable investment activity. The two steps of this stage are, first, learning to provide a turnkey plant embodying a given technology, and, second, learning to innovate completely new processes or products. The last step is what we normally regard as a major innovation and requires 'basic' R & D on the frontiers of particular technologies; something which would usually be confined to the developed countries. However, the line between minor improvements and major ones is very difficult to draw, and developing countries may be on the brink of entering this stage in certain sectors. Certainly, India has launched some 'basic' R & D programmes which have yielded processes novel enough to be placed in this category[26] – though they are not 'breakthroughs' in the Schumpeterian sense.

This threefold classification of technical learning is clearly oversimplified and the lines between the stages are difficult to draw, but it highlights the main steps involved in the emergence of developing countries as exporters of technology. Different sorts of technical learning give rise to the ability to export different forms of technology. The various learning processes can take place in a number of different sorts of firms within the country, which would then have to come together to achieve different kinds of technology transfer. It is also possible, and indeed our evidence shows that it is quite common, for missing bits of 'learning' to be supplied from abroad, and for domestic firms to concentrate on certain parts of the technology where their own learning is advanced. Bearing these complexities in mind, let us see how far we can identify the stages of learning with the nature of technology exports.

Stage (1), or 'elementary' learning in the terminology of my earlier (1979) paper, can give rise to two forms of technology exports: direct investment and licensing. Direct investment can exploit the second step of stage (1), managerial and technical mastery over a given technology, with all the equipment and basic designs being imported from third countries. Licensing can be used to sell particular improvements which have been made to imported processes. By definition, this would apply to industries that do not produce capital goods.

Stage (2), or 'intermediate' learning, can give rise to direct investment and licensing in capital goods industries, as well as certain types of consultancy exports for detailed engineering or the implementation of certain parts of production technologies.

Stage (3), or 'advanced' learning, can give rise to direct investment, licensing in the broadest sense, turnkey activity, and specialized or general engineering consultancy (depending on the range of capital goods technologies experienced).[27]

As one goes up the learning scale, the scope, complexity and scale of potential technology exports rise, as does the commodity export content (machinery and intermediate goods supplied) of the disembodied exports. Whether or not the potential is realized, however, depends on whether the non-technical aspects of learning have kept pace with the technical aspects, and, more fundamentally, on whether the learnt technology is internationally competitive.

b. Non-technical Learning

This term refers to the whole gamut of functions, from organizational, managerial, financial and marketing to political ones, involved in successful commercial activity. Some of these functions are learnt along with technical learning, but others may develop (or not) independently. The failure of the latter to evolve may hold back or abort technical learning; their rapid growth may lead enterprises to export entrepreneurship *per se* rather than technical ability, or to speed up the pace of technical learning. The distinction between technical and other aspects of learning should not, however, be drawn too sharply. The two are so intertwined and interdependent that to separate them at all may be misleading in many cases.

Given that 'learning' processes generally have led to the successful absorption of technology,[28] there may be circumstances in which this does not lead to exports of technology. Some of these are as follows:

i. The process may be totally unsealeable abroad because it has been replaced by new technologies that are more efficient at all possible factor-price configurations (e.g. in chemicals).

ii. The product may be so specialized or out-of-date that insufficient demand exists for it abroad (e.g. old products in electronics).
iii. Both product and process may be viable, but the cost of technology export may be too high (because domestic equipment is too expensive), or the lack of an established brand name may be a crippling barrier.
iv. The enterprises in which the learning is embodied may not wish to venture overseas.
v. The enterprises may be too small individually to spare the finance or personnel to compete internationally.
vi. The recipient or buyer may also be too small or insufficiently informed to seek foreign technology (as with village industries).

It follows from this reasoning that technology exports by developing countries will be undertaken in industries characterized by relatively stable technologies (where older processes and products are competitive), by the achievement of fairly competitive prices for the equipment supplied (assuming that personnel can be supplied cheaply, and to some extent outweigh higher equipment costs), by large, outward-looking firms or by agencies which co-ordinate the actions of small firms. Furthermore, we may expect to find that developing country exporters do better in international markets where the buyer is 'well informed' and can evaluate the merits of competing suppliers on cost and technical grounds rather than on the basis mainly of brand names. This is supported by the fact that nearly all Indian sales of turnkey plants have taken place in open international tenders, with a 'well informed' authority acting on these grounds.

Sufficient evidence does not, unfortunately, exist for us to test any of these propositions rigorously, but the impression conveyed by the evidence on actual exports, patchy though it is, is in conformity with our reasoning. No doubt more detailed research will reveal some other factors which promote or retard success in international markets.

What seems incontrovertible is, however, that technology exports are only the tip of the iceberg of technological activity that is going on in developing countries. If the understanding of such activity is the real object of interest, then a study of technology exports is useful only in that it is one, relatively easy to monitor, index of technological progress. There may be other, more direct, measures. However, a study of technology exports offers two advantages: First, the fact that exports have taken place implies that local technological capability is internationally competitive and has not been wasted on unviable technologies or products: with a study of other measures of technical progress (e.g. value added per employee by industry) this assumption cannot be made. Second, it is easier to study the characteristics of technology produced by developing countries when it is compared in an international market alongside technology offered by the developed countries. In particular, a carefully matched set of projects undertaken by developed and developing countries can yield valuable insights into the nature of technical change within each group.

IV A CLASSIFICATION OF EXPORTED TECHNOLOGIES

As we just remarked, the peculiar characteristics of technologies exported by developing countries can be assessed only after a detailed comparison with those provided by developed ones. In the absence of such comparisons, we can only advance some presumptions, based upon the scattered data presented in earlier parts of this study. These data suggest a threefold classification of Indian technology exports *vis à vis* technology provided by developed countries.

1. Competitive technologies, where Indian enterprises are able to provide an entire package in direct competition with developed countries, to identical standard of technical performance;
2. Complementary technologies, where Indian enterprises are able to provide the simpler, more 'labour intensive' parts of technologies to complement the more skill-intensive R & D based technologies provided by developed country firms; and
3. Non-overlapping technologies, where Indian enterprises provide products or processes no longer sold by developed country enterprises, because they are too old-fashioned, labour-intensive or unsuitable for the conditions of the rich economies where they develop their know-how.

Examples of each are easy to find. The sale of power stations by BHEL or other types of modern plant by other Indian enterprises are instances of directly competitive technologies. The subcontracting of project construction, supervision or detailed engineering by Indian enterprises from developed country firms illustrates the sale of complementary techniques. And the export of scooter technology by Bajaj, based on technology which is being phased out of Italy, is an example of non-overlapping technology; another is the export of village industry technology.

This classification should not lead us to think that only directly competitive exports affect the sales of technology by developed countries. On the contrary, both complementary and non-overlapping technology exports can reduce the market for developed country enterprises, the former by taking over the relatively easy parts of technological work which developing countries can perform more cheaply, and the latter by reducing the need for the most modern processes or products. The labels are attached purely for convenience. They refer to different parts of a continuous spectrum of technologies from the most complex to the least. At the most complex end, developing country enterprises, if they can enter at all, can only perform a few complementary services. At the medium range they can be directly competitive, and at the simplest end they take over technologies which are clearly unviable for developed country firms or which are irrelevant for them.

In the light of these considerations, what are the advantages to technology importers of buying from developing countries? First, for competitive technologies, they can benefit by getting the product cheaper, both because skilled labour is relatively inexpensive in developing countries and because the entry of developing country enterprises introduces new competition into oligopolistic markets. For complementary technologies, similarly, certain parts of the transfer process become cheaper (though it is possible that the main contractor, the developed country firm, pockets the saving).

Second, for non-overlapping technologies, the recipient benefits from the 'intermediate' nature of the product or process. The various advantages of intermediate technologies have been amply discussed in the literature and need not be repeated here.[29]

Third, developing country enterprises may be more adept at setting up plant, developing suppliers, imparting training and bringing the product to market in the conditions of other developing countries than firms from advanced countries. In other words, the experience of having absorbed technology in their relatively primitive home environment should make it easier for the developing country firm to effect the transfer than for a developed country firm: the transfer should take less time, less cost and take better root. This inference may be supported indirectly by evidence that there may be high costs of transferring technology between countries even *within* a multinational company,[30] with the costs depending, among other things, on the familiarity of the enterprise with the transferee's environment. Thus, a developing country turnkey plant builder or engineering consultant will find its home experience of greater relevance than one from an advanced country.

Finally, developing country enterprises may provide the technology in a more 'unpackaged' form than those from advanced countries, since they do not enjoy monopoly power based on the combination of advanced technology, brand name, financial resources and managerial resources that give rise to multinational companies in rich countries. The sequence of the form that technology exports have taken for a number of Indian enterprises has been the opposite of that observed for the established multinationals. Instead of entering a country with a fully 'packaged' technology in the form of a wholly-owned subsidiary which is then slowly and reluctantly unpackaged, Indian enterprises have usually started by exporting and, reluctantly, moved to turnkey plant construction and/or direct investment at the request of the host country.

This is not to argue that developing country enterprises are better 'corporate citizens' or more concerned with host country welfare than the traditional multinationals, simply that they do not *have* the monopolistic advantages enjoyed by the latter. However, it may also be true that they are more sensitive to the needs for local participation, local technical development or local 'linkages' with suppliers, than the real multinationals. Certainly, there are several official Indian statements to this effect: only further research can show how seriously they are taken.

10 Determinants of Technological Development

In this chapter we shall consider some of the general factors which determine the process of technological development in less-developed countries, which, in other words, affect how far progress is made along the different stages of the learning sequence and how widely this progress is spread over different industries. We shall also advance some tentative hypotheses as to why India, with its low per capita income, slow rate of growth and a rather faltering export performance, nevertheless it seems to have emerged as the leading exporter of technology in the Third World. We shall not at this stage try to judge whether or not success in technology exports has been worth the cost and effort that have gone into generating the technological developments that underlie it. Insofar as such a judgment can be made, we shall leave it for the following chapter.

I DETERMINING FACTORS

The factors that determine the pace of technological development are a mixture of 'given' economic conditions and official policies, with the two closely intertwined in most cases. Let us start with those which are least policy determined and work up to those which are most so.

Size of market

The size of the domestic market is of crucial significance in affecting the nature of technological development, because it determines the extent to which a diverse range of capital goods industries can be successfully established. It has been argued above that *every* sort of production activity creates some sort of technological capability, from the smallest of village industries to the most complex of modern machine building. However, the potential for learning beyond small adaptations to imported capital equipment can only be realized if there is a domestic capital goods production base. Rosenberg (1976) and Stewart (1977), among a number of other scholars, have emphasized the crucial role of machine building (expecially machine tools building) in the generation and diffusion of technical change.[1] Clearly, the size of the home market, depending upon such given conditions as size of country, per capita incomes, income distribution and the like, will be a factor of prime importance. Exporting does, of course, offer the possibility of undertaking capital goods production (with large inherent scale economies) even in small markets, but the complexity of basic capital goods design and engineering means that production for export will be either confined to simple labour-intensive operations within an international framework of production by MNCs (as with electronics) or else be conducted by MNCs or local licensees with designs entirely imported from abroad. While this may be conducive to economic growth for other reasons (as we have witnessed for a number of the newly industrializing countries),[2] it will not give the economy a base for the export of the entire technology involved. For most complex technologies a period of domestic production is required if the *entire* technology is to be mastered.[3]

Skill availability

The importance of having sufficient numbers of technically trained personnel who can then transform problems of production and application into feasible solutions is so obvious that it hardly needs stressing. We need not dwell on this at greater length, for it takes us into broad fields of education, science and training policy which are well-known in their own right.[4]

Protection of capital-goods production

A large domestic market merely permits efficient capital goods production to take place – it is not a sufficient condition for it to be actually undertaken. In most circumstances a period of protection will be required if the complex and novel activities involved in making capital goods are to be learnt and established in a developing country (we exclude from this 'offshore assembly' activity by MNCs). Such protection could take several forms. The most common (and probably the most effective) is simply to ban competing imports, but there are less drastic measures like flexible tariff protection, preference in buying on behalf of major purchasers, special requirements which discriminate against foreign manufacturers, and so on. Or protection may be substituted by subsidy in a variety of forms. Whatever the policy, there may be strong grounds for advocating 'infant industry' protection for capital goods manufacture.[5]

Promotion of local R & D

The explicit encouragement of technological work by a government can take two general forms: the creation of a research infrastructure not directly related to the production system, and the encouragement of R & D within production enterprises themselves. The Indian government has gone into both, with special emphasis on the setting up of a large number (34) of national research laboratories under the Council of Scientific and Industrial Research (CSIR) responsible for developing a large variety of industrial and agricultural technologies (with the great bulk of expenditure going to atomic, space and electronic technology), and on the commercialization of domestic innovations by the National Research Development Council (NRDC) and the Inventions Promotions Board.[6] Table 10.1 below shows total R & D expenditures in India in the public and private sectors, up to 1979, and the numbers of scientific and technical personnel employed in formal R & D activity.

TABLE 10.1 Research and development expenditures and employment in India

Year	*R & D Exp. (Rs. million)*	*R & D as % of GNP*	*Total scientific and technical employment in R & D*
1958/9	290	0.23	20,724
1968	1310	0.44	73,634
1971/2	2140	0.54	103,767
1978/9	4000	0.60	n.a.

SOURCES
NCST (1973) and *Economic and Commercial Weekly* (4 November 1978).

The table shows that the formal R & D effort has grown rapidly in the last two decades, increasing over thirteenfold in terms of expenditure (in current rupees) and over fivefold in terms of the technical personnel employed, and nearly trebling its share of GNP. While the effort is small in relation to developed countries, in absolute terms it is quite large, and not unimpressive for a developing country.

Formal R & D figures may, however, be misleading for two opposing reasons. First, they may overstate the amount of technological work that is applied to production. A large part of the money may be spent on scientific work that is unrelated to the production system and that is not commercially useful, given the gaps that Cooper (1974) has well described between the scientific and industrial establishments of developing countries.[7] Second, they may understate technological work greatly. They do not take account of such work undertaken outside formal R & D institutions within the production enterprises in the normal course of their investment and manufacturing activity. As we have argued above, and as Katz (various) has documented in detail for Argentina, this is the chief source of technical progress in developing countries. Given that their development in technology is bound to be mainly of the 'minor' variety, of assimilating and adapting technologies developed abroad, technical progress can be quite rapid even without investments in 'basic' scientific and research work, and with fairly modest investments in applied development work.

As for the promotion of in-plant R & D, the Indian government offers various fiscal incentives to enterprises for this purpose. These range from grants and tax concessions to special privileges in licensing and in importing equipment and maintaining contacts abroad. The Department of Science and Technology has a scheme for 'recognizing' in-house R & D in industry, which had, by the end of 1977, granted recognized status to 54 public sector and 430 private sector manufacturing enterprises' research establishments. Of the private sector units, 349 were medium to large scale firms. The sectoral distribution of the recognized units was: chemicals 33 per cent; electrical and electronic 28 per cent; engineering 30 per cent; other 19 per cent. They spent Rs. 400 million (about $50 million) on R & D, or 1 per cent of their turnover, and their R & D manpower came to about 10,000 persons.[8]

Several observers, notably Bhagwati and Srinivasan (1975) and Desai (1980), have commented on the very rapid increase of industrial R & D in India, and have also noted that much of private R & D is highly concentrated in a few large industrial houses. Size seems to be a crucial factor permitting substantive research effort to be launched: to return to an earlier point, the government's policy of holding back the large houses in the interests of equity and competition have probably served to retard private sector innovation. In-plant R & D, whether public or private sector, is, however, a relatively small proportion of total R & D in the country. Upadhyaya (1978) notes that, in 1971/2, it comprised only 7 per cent of total formal science and technology expenditure in the country, as compared, say, to 75 per cent in the US and 65 per cent in the UK and Japan. While there may be a case for the government taking a lead in this activity in a poor developing country, the figures suggest strongly that much of the official Indian effort is going into 'prestige' projects (space and electronics) which do not yield much tangible benefit to the productive system. However, we need more evidence before we pass any final judgement on the efficacy and productivity of such research, and the controversy which is currently raging in India on this issue does not suggest that any easy answers are possible.

Table 10.2 sets out some estimates made by Desai (1980) for the value and distribution of *industrial* (as opposed to total) R & D in India for the period 1950–74. These data are far from satisfactory: they do not include research done by engineering consultants or in the defence establishments; the figures

TABLE 10.2 Industrial R & D expenditures in India, 1958–74 (Rs. million)

	1958	*1965*	*1968*	*1970*	*1974*	*% in 1974*
Central ministries and government enterprises	—	17	22	38	245	29.4
CSIR[a]	28	77	102	117	194	23.3
Industry associations	—	6	9	11	28	3.4
Private enterprises	2	18	90	131	367	44.0
Total	30	118	223	300	834	100

NOTE
[a] Council for Scientific and Industrial Research. The data given here pertain only to expenditures of laboratories doing industrial research.
SOURCE
Desai (1980).

for the CSIR seem to be overestimates because they may include administrative expenditures; many private firms may not have reported their R & D, and different reporting enterprises may have used different definitions of R & D (some may have included 'troubleshooting' and technical expenditures and others not). Despite these deficiencies, the data are useful in illustrating the trends and distribution of industrial research.

Private sector firms dominate industrial R & D, despite a possible under-reporting on their part, and have shown a remarkable increase in the past 16 years. Central ministries and public sector enterprises (individually accounting for Rs. 100 million and Rs. 145 million respectively in 1974) come next; if public and private industrial enterprises are taken together (Rs. 512 million) the share of in-house research by manufacturers comes to 61.4 per cent of the total, a share which has risen dramatically from under 30 per cent in 1965. If data on in-house research by engineering consultants which have split off from capital goods manufacturers were available, this share would rise even further. By all measures, therefore, the data support a contention that purposeful technological activity in India is vigorous and expanding rapidly. As noted above, however, this activity is dwarfed by official research programmes in non-industrial fields.

Table A.7 sets out the industrial distribution of R & D as reported by the CSIR. It is interesting to note that a number of 'high technology' sectors in advanced countries (drugs, instruments) are also high R & D spenders in India, while others (electronics) are not. The determinants of such expenditure are clearly different, but it would need much more detailed research to identify the precise causes for this. In any case, the level of R & D spending, while not high in relation to sales, is still rather surprising for a developing country, higher than what *a priori* notions (at least the present author's) would have led one to expect.

Protection of local learning

While the recent expansion of R & D, together with large markets and a sizeable population of skilled manpower, contribute towards the explanation of Indian technological development, we still have to explain *why* the growth in R & D expenditures took place at all. Why, in other words, should industrial enterprises have undertaken the uncertainties, costs, delays and risks[9] that are inherent in R & D activity, if more highly developed, tested and commercially proven technologies were available on world markets? There are two possible answers, between which we have to choose: first, that local needs and capabilities were such that it was obvious to the enterprises that independent technological work was profitable, and that foreign technology would be too expensive or inappropriate. Thus, it was a deliberate, profit-motivated choice in favour of indigenous over foreign technology. Second, that enterprises would in fact have opted for foreign technology if they had been given a free choice, but that the government forced them to undertake their own R & D by restricting the inflow of foreign technology, keeping out MNCs in sectors where local technology was available, and protecting the markets of domestic enterprises which developed their own technology.

It is not easy to choose the correct answer without qualification, but the evidence points to the second rather than first in the case of India. It is plausible, in other words, that to undertake advanced technical activity, Indian enterprises would only have been prepared to bear the cost and risk if their 'learning' were *somehow protected against the threat posed by foreign technologies.* Some adaptive technological activity may certainly be undertaken even if foreign technology were freely available, but it is doubtful whether more basic design and development work, where uncertainty and risk are higher, would be launched by technically inexperienced, relatively new, developing country enterprises.

This is, of course, the classic 'infant industry' argument, but applied to the production of technology rather than only of commodities. It entails *two forms* of protection: of the manufacturing process itself, and of the learning process. The term 'protection' is used widely here to include not just tariffs or similar forms of reducing competition directly, but also such intervention as subsidies, concessions, political support, and the formation of monopolies.[10]

The manufacturing process needs, if the conditions for infant-industry protection apply, to be shielded for some time against imports which are cheaper because they enjoy scale economies or the benefit of elementary 'learning by doing' by virtue of having started earlier. It is presumed that local production will be competitive in a short enough time to make the present value of the future benefits (discounted at an appropriate rate) larger than the costs of protection.

The learning process, similarly, needs protection against the import of improvements in technology – the initial input of technology will generally be from abroad – so that some of the more severe risks of building up a technological capability are reduced. This is justifiable only when the discounted value of costs, not just of R & D expenditures but also of output lost, higher prices, lower quality and so on, is lower over a period than that of the benefits yielded by higher rates of growth, more appropriate techniques and products, greater exports, more use of local materials, and several forms of externalities. The calculation of costs and benefits should lead to a selection of certain types of technological work which are worth undertaking locally and not of others which need to continue being imported.

The two forms of protection need not go together. Domestic production may be protected against import competition while its technology is continuously replenished from abroad. It is also possible, though this is less likely in developing countries, that production is left unprotected, but domestic technological effort is subsidized and foreign technology is excluded. If technological development is protected, it is more likely to *entail some protection of the output produced by its results*: the two must to some extent go together if basic R & D is to be undertaken in developing countries.[11] Note that protection of the domestic market may well (as it did in Japan) *go together with an aggressive export strategy*. What, however, is the sort of continuous technology import against which local learning may need protection?

There are two main avenues for continuously importing 'ready-made' technology from abroad: *direct investment* by foreign enterprises and the *licensing* of local enterprises (and intermediate positions such as joint ventures). The direct entry of wholly owned MNC affiliates probably provides the most powerful means of importing a continuous stream of ready-made techniques and products based on the frontiers of technological work abroad. Licensing provides a channel of importing particular technologies for specified periods, and so is much more limited. Joint ventures lie between the two, the strength of links with foreign technology depending on the nature of participation, the complexity of the technology and the relative capabilities of the two parties.

The import of technology in these forms can be said to stifle the growth of local technological capability only if two conditions are met: first, the technology is such that local efforts would lead to an internationally competitive (and/or socially desirable) technology; and, second, that the technology importer does not himself undertake the investments locally to build up this capability. It may happen, for instance, that a multinational affiliate *has* to undertake local R & D work for a number of possible reasons:

i. The nature of the product – some products need modifications to suit local tastes (food or cosmetics), local regulation (medicines), local conditions (vehicles) or specific needs (specialized machinery). For these, a certain amount of local R & D is essential to adapt the basic technologies developed abroad, and for this sort of technical work, MNCs may well take the lead.
ii. Material availability – the raw materials available locally may be different, or import substitution policies may force manufacturers to develop substitutes or to help build up local supplies. All these would affect MNCs or licensees just as much as local firms.
iii. Size of local market – the market may be large enough, or the country may be used as a base to export to a sufficiently large market in the developing world, to make it economical for the MNC to invest in substantial product development.
iv. Cheap scientific personnel – it is possible, though there is little evidence that this has been an important factor, that the availability of cheap manpower would induce MNCs to base their R & D work in developing countries.[12]

If the MNC's R & D is sufficiently large, it may well generate as much technological activity as a local firm, cut off from access to foreign technology, would have done. Even more, it may well generate it more efficiently, with more resources and with better scientific backup from abroad.

It may, however, happen that the multinational affiliate (or joint venture or licensee) remains passively dependent on its foreign partner for all improvements to technology, or for all technological work beyond the level of minor adaptation or detailed engineering. In this case, a viable technical capability to assimilate, improve and export the entire technology will not be built up in the developing country: the particular enterprise concerned will not find it economical to undertake the cost and risk of reproducing technological work already done (and proved) abroad. Neither will its local rivals, since they will not be able to compete on open markets if they try to build up their own independent technology. Any firm which is to remain competitive from the start would have to resort to foreign technology (unless the industry was such that foreign technologies were totally irrelevant, e.g. cottage industry).

For the foreign producer of technology, there are significant economies of scale and of communication as well as various externalities in centralizing its basic design and development work at home.[13] It is clearly not profitable for it to set up similar activities in every area of affiliate operation, especially to support a relatively small developing country operation. The available data on MNC R & D abroad support this reasoning: there is very little of it in developing countries, and what exists is devoted to minor adaptive work.[14] Given these tendencies, basic design and development work, especially in the pivotal sector of capital goods production, would tend to be retained in the advanced countries, even for technologies that were relatively stable and could economically be transferred.[15] In these cases, learning in developing countries may remain at the elementary stage or at the imitative (detailed engineering) phase of the intermediate stage. The most difficult and risky type of learning will not be experienced in the developing country.

It is in order to progress to the higher stages of technological learning that protection of local efforts is required. Elementary learning goes on in every productive enterprise, regardless of the origin and development of the basic technology involved. More advanced learning, of the basic engineering and technical principles involved, can only take place if the learning enterprise is given an assured market and protection against the import of ready-made and riskless technologies from abroad. In general this will require a judicious restriction of MNC entry and of other forms of easy access to foreign technology combined with a judicious use of import protection. Needless to say, this is *not* an argument for seeking to do everything in every industry: here, as in trade, the principle of comparative advantage applies. And, to reiterate an earlier point, such a strategy is perfectly compatible with a strong outward-looking policy for the promotion of exports.

As noted above, the protection of local learning can take several forms. Certain activities are protected naturally in that foreign technologies are irrelevant (e.g. for handicrafts or certain agricultural products or techniques); where they are so diffuse that MNCs do not have a stronghold or are not active in them (simple manufactures of food, metal, generic drugs); or where they are not easily imported via MNCs (civil construction in many developing countries is not open to foreign building firms). Others are protected for strategic or political reasons rather than to promote local learning (communications, armaments, transportation). And others may be protected, to build up national control over production and technology, by the following:

keeping foreign enterprises out of specified sectors altogether;
limiting foreign participation in some sectors;
screening the licensing process, limiting the types and periods of agreements and prohibiting licences for technologies which can be developed locally;
subsidizing local firms engaged in technological work, helping their efforts by negotiating licenses on their behalf, or directly providing them with the results of official technological work;
encouraging the growth of very large local firms which have strong political and financial backing, and which can thus protect their own technical development, and transfer it across industries; and
investing directly in R & D facilities.

Several different countries have, at different stages, adopted different policies for protecting local learning. Japan furnishes an extremely successful example of a strong policy of protecting and subsidizing local technology together with the encouragement of large firms, the original *sogo shosha*. India tries to restrain its large private sector firms but has policies to force the pace of local learning by restricting foreign technology imports in a variety of ways and by promoting local efforts within its Science and Technology Plan. Other advanced developing countries are starting to intervene in the technology import process, within the framework of their own Science and Technology Plans (Brazil and Mexico) or by restricting or regulating licensing (Korea) and generally encouraging local enterprise.

We have dwelt on this last factor – the promotion of local learning – at such length because it is relatively ignored in the literature and because it needs to be distinguished from the protection of production, in theory if not in practice. We do not argue that it is *more* important than the other influences on technological development, nor do we suggest that the protection of learning is not (as with all forms of protection) fraught with costs and dangers. The intention is more to advance an interesting hypothesis which may be of relevance to important aspects of theorizing and policy making. Let us, then, see how we can explain India's relative role as a technology exporter in the light of these considerations.

II REVEALED COMPARATIVE ADVANTAGE IN THIRD WORLD TECHNOLOGY EXPORTS

We have established that a number of semi-industrialized developing countries have entered international technology markets as successful exporters of a variety of industrial technologies in several different forms. We have suggested, but the final proof must await more research, that India is the leading exporter of industrial technology in terms of sophistication, range and diversity of its technology exports. In comparison with other semi-industrialized countries, however, India has not performed particularly well in terms of growth of GNP or exports. It has a long experience of industrial activity and a large capital goods sector, but so does Brazil, and its industrial production is more than double that of India's. It has a well-developed local industrial class, but Argentina has a strong local entrepreneurial class also, as have Korea and Taiwan. All these countries have a strong base of engineering education.

One possible explanation is that India has a larger supply of skilled technicians and engineers, and that the cost of these is much lower than for competing developing countries. No doubt sheer technical availability[16] has helped greatly with building up India's capabilities, but it is not academic degrees that lead to technology exports – it is practical experience with designing and implementing technologies. If the supply of engineers had not been large, the extent of learning would have been much less; however, if the conditions for learning had not been there, the availability of engineers would have been irrelevant. The cost element is not an important consideration in explaining Indian success *vis à vis* other developing countries. For a large range of technologies India has had to compete against developed countries – not developing ones.

Another possible explanation is the amount of money invested by the government in the research laboratories – perhaps a unique experiment in generating 'basic' research in a poor country.[17] The Fifth Plan envisages an expenditure of Rs. 23 billion ($2.9 billion) on science and technology. A number of processes developed there and commercialized by the NRDC are now in use, most of them in the small-scale industrial sector.[18] It is doubtful, however, that this has contributed greatly to India's success in technology exports.[19] Only a few of the processes held by the NRDC have been licensed abroad (see Table A.5), and few of the major technology exporters are on the list of enterprises acting through the NRDC. The know-how on which technology exports are based is mainly concerned with the construction, design and engineering of large industrial plants; the know-how produced by the scientific infrastructure set up by the Indian government is heavily weighted by atomic and space efforts. It has led, in the manufacturing field, to relatively few specific processes, and not many of these are in the fields in which India has exported technology. This does not necessarily mean that the scientific establishment's contribution in other ways is negligible, merely that their technical progress has been localized in agents and forms which do not lead to technology exports.[20] Despite the small proportion of formal R & D conducted in manufacturing enterprises, it is their technical activity and experience more broadly which has contributed to the country's success in foreign technology sales.

We are left, then, with what we have broadly termed 'protection of local learning' in the manufacturing enterprises as the main explanation of India's success in technology markets.[21] India is the only one of the semi-industrialized developing countries which has, consistently and over a long period, pursued a policy of building up an indigenous technical capability, seeking in this, as in production, as much 'self-reliance' as it could possibly get.[22] It may well be that this pursuit has in many instances been wasteful or misdirected, but at the same time it has encouraged, cajoled or forced local enterprises to develop their technology. In particular, the close restrictions placed on technology imports by MNC entry has protected basic learning in several high skill and complex industries in the capital goods sector. These are precisely the industries which can act as focal points for technical progress in a broad range of user industries.

Technical learning has progressed in both foreign and domestic-owned enterprises in the manufacturing sector. In some cases, circumstances such as those mentioned above may have induced MNC affiliates to play a leading role in local innovation: the largest private formal R & D establishment in the country belongs to Hindustan Lever, Unilever's subsidiary. More generally, however, local enterprises seem to have played the leading role in assimilating and exporting technology in the investment goods sectors. The evidence clearly indicates that technology exports have been conducted, with only a few exceptions, by locally-controlled firms.

This could imply one of two things:

a. Given appropriate conditions for local learning, a local firm will progress further up the learning ladder than a multinational affiliate; or
b. both local and foreign firms progress equally, but a local firm would gain more from exploiting its knowledge by exporting technology than a foreign affiliate. A foreign affiliate may simply have repeated the learning possessed by its parent organization, in which case the exigencies of global planning by the MNC may well restrict it from selling its technology and so competing with other parts of the company. Or it may have learnt something new, in which case the commercial exploitation of the innovation may be undertaken by the parent organization,[23] either because this is more in conformity with its global strategy, or because it is quicker and more efficient (there are learning costs involved in tapping foreign markets, and the developing country affiliate may not be allowed to incur these when foreign marketing and manufacturing networks already exist). Thus, technical progress will be internationalized by the multinational company rather than by the affiliate in the developing country. These two may coincide, of course, where the local affiliate is large and is used as a major base for serving other parts of the organization (to conduct R & D and to sell technology); in most developing countries, the two will tend to be different.

A comparison with technology exports by other countries appears to confirm the hypothesis about the relationship between revealed comparative advantage in technology exports and the protection of local learning. Countries which have large industrial sectors, long experience of industrialization and fairly outward looking policies (the big Latin American countries) have been able to achieve technology exports mainly in sectors where MNCs have not dominated their industries. Brazil and Mexico, for instance, export mainly petrochemical and steel technology, where the state has intervened to protect local enterprises, by setting up government-owned enterprises (PEMEX and PETROBRAS) or by subsidizing private local enterprises. Argentinian enterprises have exported technologies mainly in relatively simple industries (food processing, metal products) where copying is easy, where the traditional processing of local materials lent an advantage, and where the large multinationals are not particularly active. Korean, Taiwanese and Hong Kong enterprises have, similarly, exported the rather simple technologies where there is the 'natural' protection given by the wide diffusion of the know-how. And exports of civil construction know-how have been fostered by the 'natural' protection given by the nature of the activity and official policy. Korean enterprises have, however, advanced somewhat further because of the close links they have fostered with their government, their relatively large size and dominant positions in local production and exports, and the heavy official promotion of all forms of export activity.

In the same large Latin American countries, in sectors where MNC entry has been allowed freely, there has been little activity by local enterprises in the category of 'fully competitive' technology exports. A lack of protection of advanced learning by local capital goods producers,[24] a rather passive dependence on imports of licensed technology,[25] or the elimination of local enterprises altogether from high-technology areas of activity have all contributed to the inability of these countries to master and export these technologies. The heavy electrical equipment sector provides an excellent case in point. In India, the setting up of Bharat Heavy Electricals Limited (BHEL) led, after years of difficult learning on both technical and organizational fronts,[26] to the emergence of a technology exporter of world class in a sector previously entirely the preserve of developed country firms. In Brazil, on the other hand, the lack of an official policy to protect and promote local enterprises (which were active some 20 years ago) has led to aggressive inroads by the leading electrical MNCs, and to the relegation of local firms to relatively minor and low-technology areas of activity.[27]

III LIMITATIONS TO TECHNOLOGICAL CAPABILITY

We have hinted several times at the limitations to technological learning by developing countries. Since local learning is necessarily based on research, design, development and investment experience at home, any technology which cannot be subjected to such experience would lie beyond the reach of developing country enterprises. The factors that would affect this division are as follows:

i. The scale of R & D required: clearly, developing countries cannot hope to reproduce the very expensive and advanced R & D which is required on the frontiers of several technologies. Thus they cannot compete with the leaders of advanced countries in most industries in products and processes which are the innovative stage. As noted previously, the distance that they lag behind the frontiers depends essentially on the speed of advance abroad and the resources needed to achieve such advance.

ii. The scale of investment needed to implement technologies: while a technology itself may not be changing very rapidly, its efficient implementation may require investments of a size which is unfeasible for developing country markets or out of the financial reach of developing country enterprises.

iii. The sophistication of markets: certain consumption goods are geared to very high income consumers and cannot be economically produced by developing country enterprises for their home markets. Thus, the technology for expensive motor cars may not be within the reach of developing countries, in contrast to that for ordinary mass-production cars or some commercial vehicles. Developing countries may, within an integrated MNC structure, well be able to produce some components for high income products, but they can never gain sufficient experience of the entire technology to become competitive in it.

iv. The nature of the process: it may be that the production technology of batch production methods is easier to master than that of large-scale continuous production (e.g. machinery as opposed to chemicals). This is due to the possibility that processes which can be broken down into several discrete parts are simple to copy once the basic engineering principles are known, and because improvements and adaptations can also be made in small steps, while to assimilate a continuous process system involves mastering the entire technology together.

These four considerations lead us to expect that developing countries will progress fastest when

(a) the technologies are relatively stable internationally;
(b) the skills required are gained through practical engineering experience rather than large-scale and expensive R & D activity;
(c) the products are not aimed at meeting the needs of high-income, rapid-change-orientated consumers; and
(d) the processes are not inherently very large-scale, and can be subjected relatively easily to adaptation, scaling down and small improvements.

It is important to reiterate that the dividing line *does not fall by industry but by particular technologies within each industry*. There will be complex, innovative technologies within the most traditional industries which are beyond the reach of developing countries; and there will be relatively standardized, diffused and easily assimilated technologies within the most sophisticated industries in which developing countries can become proficient.[28]

There are lessons for students of trade theory here. Most recent empirical examinations of comparative advantage in manufactured trade of developed and developing countries[29] have identified the competitive edge of the former as lying in high-technology and high-skill activities, and of the latter as lying in standardized-technology, low-skill activities (physical capital intensity seems generally to have been discarded as a significant influence). In many studies, high skill and high technology are regarded as practically synonymous, and both are treated as being the preserve of high income countries.

As examination of technology exports by developing countries suggests that, with the natural accumulation of skills in the process of industrialization, if furthered by appropriate technological policies, the comparative advantage of developed versus developing countries will be determined, not so much by skill requirements *in general*, but by skill inputs based on *specific learning processes which cannot be replicated in developing countries*. The four conditions we have just mentioned, for instance, will determine, even for highly skill-intensive activities, which can most economically be performed in rich or poor countries.

This will lead to a change in the specializations of developed and the more industrialized developing countries: rather than concentrating on different types of industries, as they have tended to do till now, they will specialize in different processes within the same industries. Developed–developing country trade will, in other words, come more and more to resemble inter-developed country trade, with the growing sway of the sorts of considerations that Grubel and Lloyd (1975) have advanced as determining 'intra-industry trade'. These include such factors as 'gaps' in technology, product differentiation and scale economies, all similar to the ones we have advanced as the limitants to learning in developing countries.

Some recent literature on manufactured exports by developing countries, as surveyed in my (1980) paper, has remarked on the rapid diversification of these exports into high-skill activities. In part this has been traced to the activities of MNCs, which have transferred the simpler parts of high-skill, high-technology processes to some developing activities. In part, however, if the technology export scene is a reliable index, it is due to the efforts of indigenous enterprises. It may be argued, of course, that such diversification, which necessarily entails investments in capital goods production and probably involves heavy learning costs, is somehow 'unnatural'. Developing countries which go in for a policy of technological development may be accused of distorting their 'natural' comparative advantage, which lies in labour-intensive, low-skill activities. This is a controversy which cannot be resolved without a much more detailed examination of the costs and benefits of 'learning' technology. Certainly a strong case can be made, on general considerations, in favour of developing countries investing in the building up of their technological capability, of the diversification of their exports into dynamic, modern activities, and of the exploitation of their human resources and accumulated experience.[30]

These considerations also point to the possibility that developing countries which have built up a strong technological capability will reveal different sorts of comparative advantage in their exports to other developing countries as opposed to more developed ones. As far as pure technology exports are concerned, we would expect developing country enterprises to sell only certain complementary services to the advanced countries, while they would sell much larger packages, with higher skill contents, to developing countries. As far as engineering exports are concerned, similarly, we would expect to find that in general they would sell simpler products (e.g. hand tools, simple machinery) to developed countries, which would need a higher degree of sophistication at the advanced end than developing countries can competitively provide,[31] than to developing ones, which would prefer the sorts of older, simpler or smaller-scale equipment provided by them. The 'sourcing' activity of MNCs, however, may impose a uniform technological level on all their exports, regardless of their destination.

To conclude, the explanations advanced in this chapter for the observed patterns of technology exports are still very tentative. However, they call into question some of the simpler analysis of technical progress in developing countries which do not distinguish between the various stages and agents of the learning progress, and identify the transfer of technology within the multinational corporation with the development of technological capability within the host country. Findlay (1978) has argued, for instance, that multinationals can be the main agents of technical progress in backward countries. Since they possess the most advanced technology and since direct investment by them is the most powerful means of transferring it, it follows, in Findlay's reasoning, that a greater multinational presence will lead to a more rapid 'catching up' of the host country with the advanced countries.

This argument may be valid as far as the transfer of production technology at the elementary levels is concerned. If we consider the ability of the host country to master the technology as a whole, and to engender its own technical progress, Findlay's argument may no longer apply. Multinationals may contribute to local technological capabilities in certain specific circumstances, but in general a strong foreign presence (or a heavy dependence on licensed technology) may inhibit the local process of learning. Foreign enterprises thus have two crucial roles to play, of providing the initial injection of new knowledge on which the host country can build, and of supplying the sorts of new technologies which cannot be mastered in the developing countries: whether this is best done in the form of wholly-owned foreign subsidiaries, joint ventures or licenses, depends on the nature of the technology and the state of development of the recipient. For the large area of technological work which lies beneath the difficult frontiers of advanced innovation, however, developing country enterprises can progress substantially on their own, and here a degree of protection may be a necessary condition for progress.

11 A Final Note on Costs and Benefits

In a study which is based on scattered evidence, the assessment of costs and benefits must be even more tentative than the empirical conclusions. The author may, therefore, be forgiven if he does not commit himself to a net weighing of these costs and benefits, but satisfied himself with an iteration of the most important relevant considerations.

As far as the *benefits* of technology exports are concerned, we may consider these at two levels. First, at the level of the exports proper, and second, at that of the broader technological development of which they are the manifestation.

The benefits of technology exports *per se* are the immediate gains in terms of foreign exchange from the embodied and disembodied sales made, and the longer-term gains in terms of possible continuing exports of intermediates, replacement parts and know-how. There are also important externalities involved. The building up of an 'image' for the quality of products and skills of a country will strengthen its exports generally in the recipient country and help it to win technology contracts elsewhere. The success of the leading firms will give others the confidence to enter international markets. And experience gained abroad will add to the fund of technical learning by the firms concerned, improving their performance both at home and in future work abroad.

The benefits of technological development more broadly are potentially so vast that it is difficult to describe them clearly. To the extent that technical progress is the motor behind growth – better use of existing materials and manpower, new products and techniques stimulating greater investment, diversification of exports – the growth of a strong technological capability will create at least some of the conditions for development. To the extent that it provides an independent ability to assess, bargain for, modify or do without foreign technology, it will improve the country's position in buying foreign technology and shaping it to its own needs. It may provide a greater ability to develop local suppliers of components, especially in the small-scale sector which may be left out of the main stream of activity if very new and unadapted technologies are utilized wholesale. And finally, strong technological capabilities in industry are bound to spill over to other sectors of the economy and benefit agriculture and services, by providing better inputs, larger markers and a supply of personnel capable of improving technologies there.

The *costs* of technology exports may also be considered at two levels. At the level of the exports themselves, the domestic resource cost of providing equipment and personnel may be higher than in alternative means of earning foreign exchange. This would be the case if technology exports were heavily subsidized or if official policies forced enterprises to undertake them regardless of their profitability. In the case of India the opportunity cost of technical personnel is unlikely to outweigh their foreign-exchange earnings, but the domestic resource cost of the capital goods involved needs to be carefully examined.

At the broader level of technological development, the costs arise indirectly from the protection of domestic production and learning involved, and directly from the expense of setting up a scientific and educational infrastructure and of starting up R & D programmes. The heaviest costs are probably the indirect ones. As with all forms of protection, that of 'learning' involves temporarily lost output, higher costs, delays and mistakes, even in successful cases, and much longer term costs in unsuccessful cases where the enterprise is unable to build up a viable technological capability. It may well be that by a sweeping policy of technological self-reliance in all sectors India has wasted resources in building up a top-heavy research structure of government laboratories and in trying to master technologies which, like advanced electronics, are inherently out of its reach for a long time to come, or which, like space, are mainly irrelevant to its economic needs. Many of these unnecessary costs, if such they are, could have been avoided while keeping the beneficial aspects of learning.

We argued earlier that the promotion of domestic technology involved two sorts of protection, that of production (against imports) and that of learning. It need hardly be stressed that all forms of protection need to be applied flexibly, and ultimately phased out. Protection of production may be a necessary condition for the promotion of technological capability, but after a certain stage a strong dose of competition in production may be necessary to stimulate the learning effort; or, given the need for temporary protection against imports, a dose of competition from foreign technology may be required to give direction to local R & D. We really do not know enough about the processes and risks involved to evaluate which policies are optimal in particular countries at particular times.

In India, it has been argued at length, and with reason, that import substitution policies have gone too far (Bhagwati and Srinivasan, 1975), and the structure of incentives has been unduly biased against export activity. A more outward-looking strategy, such as the one being followed by Brazil or Korea, may well have led to greater exports and a faster rate of industrial growth. However, India has registered a relatively high degree of indigenous technological development. Must we argue, on this basis, that the technological development which has taken place in India has been a beneficial (and perhaps relatively small) by-product of an otherwise wasteful strategy? Or could India have achieved the same technological advance within a more outward looking less interventionist strategy?

This is a very general and complex question to which any answer is bound to be speculative. *Some* protection was probably a necessary measure to launch many of the more complex manufacturing industries. It is possible that the initial phases could have been completed, not by giving tariff protection, but by some form of subsidy or the formation of monopolistic structures, but let us ignore this alternative here. A great deal of protection against imports probably exceeded what was strictly necessary on static infant industry grounds. Some of this 'excess' protection undoubtedly contributed to the successful building up of technological capabilities, but without the constant prodding of the government to achieve technological 'independence' (an attractive but unattainable goal) it would probably have come to nought. And some protection probably exceeded this and contributed only to a sheltered, profitable existence devoted to serving the home market.

It should, therefore, be possible for India now to pursue a much more outward-looking strategy while continuing to build up an independent technological capability. And this is not an argument for doing without a continuous inflow of foreign technology into activities where Indian enterprises do not have a comparative advantage. The costs of the import-substitution strategy were, in our view, only partly necessary to India's technological growth. With a more liberal strategy, and perhaps with some necessary modifications to its licensing and other control measures, India may well have been able to 'do a Japan', which, after all, provides the prime example of successful protection of technological learning.

We end on this inconclusive note. The present author's belief, and this is perhaps only an article of faith, is that India's technological policy, insofar as we can separate it from import substitution more generally, was on the whole worthwhile. It has been carried too far in certain instances, but its benefits are being vitiated by the heavy and continuing influence of inward-looking and excessively regulatory policies. If these policies continue to be changed, the prospects for technology exports are bright. But more, much more, research is needed into the dark ways of technological growth in developing countries before any conclusive judgement can be made about these matters. This book is perhaps more a plea for further research than anything else.

Appendix

TABLE A.1 Export of Indian engineering goods to major
(all commodities) thereto

Major countries	(1) India's total export in 1974/75	(2) Export of engng. goods in 1974/75	(3) % Contribution of engng. goods to total export in 1974/75	(4) India's total export in 1975/76
UAE	44.17	14.54	33	64.80
Kuwait	38.07	11.14	29	45.36
Iran	214.46	16.85	8	270.77
US	375.03	22.95	6	505.49
USSR	418.12	10.95	3	412.78
Bangladesh	42.18	13.04	31	62.12
Saudi Arabia	35.42	10.41	29	59.79
Iraq	72.30	38.33	53	64.54
Germany – FDR	105.04	10.25	10	117.32
Tanzania	9.16	4.52	49	16.56
UK	306.34	12.82	4	402.02
Nigeria	21.61	13.35	62	37.02
Sri Lanka	26.76	7.09	26	23.00
Indonesia	50.90	9.26	18	52.02
Singapore	36.42	9.29	25	51.04
Holland	70.92	1.91	3	75.00
ARE	52.44	4.20	8	100.29
Malaysia	28.08	17.92	64	32.32
Thailand	12.32	8.12	66	15.78
Yemen Arab Rep.	16.09	0.73	5	16.17
Oman	16.21	5.81	36	18.82
Nepal	42.41	7.54	18	50.35
Kenya	14.84	7.63	51	15.54
Bahrein	9.33	2.24	24	16.67
Yugoslavia	29.69	6.19	21	28.47
Japan	294.91	3.84	1	426.33
Zambia	9.43	4.29	45	5.47
Philippines	3.97	2.46	62	11.54
Australia	61.19	6.26	10	47.65
Hong Kong	27.19	4.60	17	40.48
France	83.90	2.20	3	83.58
Jordan	15.41	1.33	9	10.69

Countries during 1974/75 to 1976/77 and India's total export (f.o.b. value in crore rupees)

(5) Export of engng. goods in 1975/76	(6) % Contribution of engng. goods to total export in 1975/76	(7) India's total export in 1976/77	(8) Export of engng. goods in 1976/77	(9) % Contribution of engng. goods to total export in 1976/77
15.77	24	160.61	46.39	29
11.88	26	112.76	38.20	34
22.47	8	144.58	28.45	20
15.54	3	547.23	27.40	5
19.28	4	440.37	27.37	6
18.58	29	54.27	26.15	48
15.51	25	74.37	25.43	34
23.83	36	46.51	19.14	41
11.97	10	223.96	18.16	8
9.38	56	23.81	17.17	72
12.38	3	508.85	16.98	3
26.74	72	24.97	16.68	67
10.63	46	38.93	15.75	40
12.04	23	60.55	18.20	22
9.56	18	57.62	12.72	22
3.04	4	185.61	12.27	6
8.61	8	90.75	11.26	12
16.62	51	29.41	10.87	37
6.63	42	25.48	9.16	36
1.74	10	44.49	8.67	19
5.79	30	29.94	8.51	28
9.27	18	50.66	8.44	17
6.72	43	17.51	7.51	43
4.53	27	23.48	7.04	30
14.15	49	49.22	6.87	14
7.08	1	538.87	6.03	1
3.08	56	9.34	5.82	62
9.53	82	22.61	5.80	26
3.99	8	64.70	5.60	9
2.46	6	72.13	5.33	7
3.03	3	160.82	4.90	3
0.85	8	10.63	4.84	46

TABLE A.1 (*Contd*)

Major countries	(1) India's total export in 1974/75	(2) Export of engng. goods in 1974/75	(3) % Contribution of engng. goods to total export in 1974/75	(4) India's total export in 1975/76
Qatar	8.60	1.39	16	9.53
Zaire	1.51	0.41	27	2.70
Libya	8.60	4.73	55	3.90
Sudan	66.46	1.38	2	36.51
Canada	43.96	2.99	7	42.36
Mauritius	6.01	1.98	33	6.78
Poland	76.91	3.13	4	84.73
Italy	52.13	1.11	2	78.53
Sweden	15.67	0.83	5	13.13
Afghanistan	14.52	1.91	13	33.94
Germany – DR	34.40	1.82	5	24.29
Syria	7.13	1.14	16	1.97
Bulgaria	17.11	0.65	4	22.92
Burma	4.62	0.62	13	8.91
Hungary	19.44	1.15	6	14.32
South Yemen – PDRY	14.29	0.85	6	7.41
Uganda	4.43	3.53	80	2.05
Guiana	0.96	0.61	64	1.17
Greece	3.09	0.67	21	6.96
New Zealand	20.73	2.10	10	12.87
Switzerland	16.24	0.48	3	51.62
Ethiopia	3.21	0.67	21	4.97
Taiwan	3.19	0.71	22	5.86
Ghana	2.45	0.82	33	1.72
Belgium	51.79	0.81	2	39.07
Sub-total	3021.20	332.12	11%	3679.35
% Share of Total	89	96	—	93
Others	277.42	16.99	6%	251.93
Total	3298.62	349.11	11%	3931.28

NOTE
Countries ranked by value of engineering exports in 1976/77. A 'crore' is 10 million.

SOURCE
EEPC, *Handbook of Export Statistics 1976–77* (Calcutta, 1978) pp. 24–7.

(5) Export of engng. goods in 1975/76	(6) % Contribution of engng. goods to total export in 1975/76	(7) India's total export in 1976/77	(8) Export of engng. goods in 1976/77	(9) % Contribution of engng. goods to total export in 1976/77
3.27	34	19.30	4.54	24
2.35	87	4.36	4.33	99
1.17	30	13.11	4.06	31
1.09	2	51.15	3.82	7
3.15	7	48.58	3.76	8
1.70	25	11.14	3.60	32
4.60	5	112.35	3.12	3
2.07	3	116.65	2.76	2
1.21	9	25.41	2.57	10
3.10	9	21.60	2.46	11
1.47	6	42.46	2.33	5
1.18	59	4.02	2.24	56
1.41	6	21.97	1.95	9
1.70	19	9.04	1.79	20
1.23	8	21.14	1.74	8
0.85	11	7.39	1.71	23
0.98	48	2.57	1.60	62
0.83	71	2.02	1.58	92
1.09	15	4.52	1.57	35
1.92	14	12.99	1.39	11
0.37	1	69.16	1.25	2
0.80	16	9.66	1.23	13
1.48	25	17.43	1.22	7
0.94	55	1.80	1.16	64
1.17	2	98.05	1.00	1
393.87	10%	4693.22	536.89	11%
96	—	94	97	—
14.35	5%	292.91	14.79	5%
408.22	10.4%	4986.13	551.68	10.7%

TABLE A.2 India: exports of engineering goods, 1972/3 to 1976/7 (Rs. million and %)

Industry	(1) 1972/3 Value (%)	(2) 1973/4 Value (%)	(3) 1974/5 Value (%)	(4) 1975/6 Value (%)	(5) 1976/7 Value (%)	(6) % Growth 1972/3 to 1976/7
A. *Industrial Plant*	114.6 (8.1)	159.6 (8.2)	312.3 (8.7)	455.3 (11.2)	431.9 (7.8)	276.9
1. Textile, Jute	34.1	39.4	162.8	223.0	151.1	343.1
2. Sugar	11.8	39.8	33.8	45.1	24.6	108.5
3. Cement	23.2	11.0	5.9	3.0	4.5	−80.6
4. Food Processing	8.5	12.9	25.6	41.3	41.4	387.6
5. Other	37.0	56.5	84.2	142.9	210.3	468.4
B. *Machinery and Tools*	203.2 (14.4)	280.7 (14.5)	531.6 (15.2)	673.1 (16.5)	910.4 (16.5)	348.0
1. Machine Tools	21.3	36.9	71.2	83.9	169.3	694.8
2. Diesel Engines & Compressors	90.6	127.9	286.1	317.0	357.1	294.2
3. Cranes, Lifts	7.2	7.8	7.6	11.2	27.2	277.8
4. Pumps	8.1	17.5	25.2	64.6	59.5	734.6
5. Hand Tools	76.0	90.6	141.5	196.4	297.3	291.2
C. *Electrical Machinery, Products*	345.2 (24.5)	379.4 (19.6)	664.6 (19.0)	826.6 (20.2)	962.4 (17.4)	178.8
1. Power Machinery Switchgear	46.0	60.5	119.9	148.7	156.0	239.1
2. Transmission Towers	48.6	24.4	51.2	73.5	89.8	84.8
3. Wires, Cables	125.7	115.4	172.5	211.8	238.2	89.5
4. Fans	25.4	24.3	67.7	72.7	111.4	338.6

5. Electronics	47.2	92.6	125.8	134.6	126.9	168.9
6. Other electrical products	23.0	34.2	70.5	109.8	135.9	490.9
7. Batteries	29.0	28.0	57.0	75.4	104.2	255.6
D. *Transport*	289.4 (20.5)	386.7 (20.0)	647.0 (18.5)	890.5 (21.8)	939.1 (17.0)	224.5
1. Complete vehicles (commercial)	89.3	61.4	165.6	244.9	313.1	250.6
2. Vehicle parts	56.8	79.2	144.1	180.2	233.1	310.4
3. Bicycles, parts	105.5	152.1	218.4	248.5	218.2	106.8
4. Wagons, coaches	34.0	59.0	99.2	172.3	112.9	232.1
5. Ships	3.8	35.0	19.7	44.6	61.8	15,263
E. *Metal Products*	386.7 (27.4)	627.7 (32.4)	1,170.2 (33.5)	1,057.9 (25.9)	2,036.7 (36.9)	422.7
1. Boilers, heat exchanges	7.6	28.5	31.2	17.9	19.2	152.6
2. Steel structures	45.9	51.9	65.9	80.7	141.3	207.8
3. Pipes, tubes	92.0	208.0	414.0	211.4	586.9	537.9
4. Bright bars	10.9	28.2	24.5	27.2	91.2	736.7
5. Ferrous hollow-ware	13.8	19.0	39.4	32.8	68.8	398.6
6. Wire ropes, products	41.2	46.5	154.4	157.0	221.0	436.4
7. Industrial fasteners	16.7	20.8	53.8	72.4	118.6	610.2
8. Sanitaryware	40.3	45.7	79.2	94.8	229.7	470.0
9. Castings, forgings	10.7	11.2	40.8	54.2	29.5	175.7
10. Other steel products	54.0	76.5	151.2	154.2	173.9	222.0
11. Non-ferrous products	53.6	91.4	115.8	155.3	356.6	565.3
F. *Other*	71.7 (5.1)	100.6 (5.2)	165.4 (4.7)	178.9 (4.4)	236.0 (4.3)	229.1
1. Heating, Cooling equipment	9.6	16.1	22.5	30.0	34.1	255.2

TABLE A.2 (*Contd*)

Industry	*(1)* *1972/3 Value (%)*	*(2)* *1973/4 Value (%)*	*(3)* *1974/5 Value (%)*	*(4)* *1974/6 Value (%)*	*(5)* *1976/7 Value (%)*	*(6)* *% Growth 1972/3 to 1976/7*
2. Sewing, Knitting machines	17.3	12.9	20.6	12.0	17.5	1.2
3. Scientific, surgical Instruments	9.4	6.0	12.0	24.7	32.2	242.6
4. Oil lamps, stoves	11.2	19.5	22.6	22.4	32.5	190.2
5. Miscellaneous	24.2	46.1	87.7	89.8	119.7	394.6
Total	1410.8	1934.7	3491.1	4082.2	5516.8	291.0
India's total exports	19708.3	25234.0	33288.3	40428.1	57430.0	161.0
Engineering exports % Total	7.2	7.7	10.5	10.1	9.6	—
%Increase over previous year: engineering exports	12.6	37.1	80.4	16.9	35.1	

SOURCE
Extracted from EEPC, *Handbook of Export Statistics 1976-77* (Calcutta, 1978) pp. 2–4.

TABLE A.3 Exports of selected items of electrical and non-electrical industrial equipment, 1974/5 to 1976/7 (Rs. Lakhs)

Item	*1974/5*	*1975/6*	*1976/7*
Electrical power machinery	944.79	1225.73	1344.75
1. Circuit breakers	18.79	31.41	22.64
2. Control-gear and switchgear	330.94	411.39	466.82
3. Generators and alternators	106.75	120.79	147.82
4. Rectifiers	4.19	2.37	2.84
5. Relays	5.35	4.78	4.60
6. Transformers	318.43	426.23	491.49
7. Torque convertors and invertors	6.56	13.01	—
8. Power machinery (n.e.s.)	153.78	215.75	208.54
Furnace and ovens	29.23	40.48	30.40
Fans and parts	676.66	726.99	1113.85
Fluorescent tubes and fixtures	262.59	292.06	160.50
Lamps and Bulbs	44.07	44.71	111.69
Measuring and controlling equipment	22.64	10.20	72.33
Electric motors, starters and pumpsets	254.35	261.40	215.70
1. Motors	195.45	192.80	160.47
2. Starters	42.90	60.44	55.23
3. Pump sets	16.00	8.16	—
Radiation equipment	—	5.42	—
Radios and parts	408.11	588.15	658.04
Relay signalling and train lighting equipment	14.50	23.61	35.41
Wireless and other electronic equipment	162.85	108.06	72.84
Machinery non-electrical	7539.31	9425.11	11322.73
Diesel engines, parts and pumpsets	1910.92	2111.17	3363.65
Mechanical pumps	252.62	645.65	595.35
Agricultural machinery	114.32	194.49	204.85
Air compressors	177.61	205.37	206.50
Air conditioners, refrigerators etc.	224.77	299.98	340.64
Asbestos cement plant and mill machinery	58.66	30.44	45.06
Ball and roller bearings	82.44	40.00	81.59
Battery manufacturing plant	7.60	—	—
Bicycle manufacturing plant	1.32	—	—
Boilers and fittings	312.27	103.44	130.61
Cement mixers and concrete vibrators	60.79	180.32	62.70
Chemical plants	15.27	91.21	188.03
Cigarette and match making machinery	5.30	68.09	52.08
Cinematographic machinery	34.56	26.53	29.37
Coffee processing machinery	3.82	0.58	6.54
Coke oven plant	—	74.64	—
Construction and mining machinery	19.32	28.36	75.58
Drilling rigs	—	106.55	35.04
Food processing machinery	0.32	2.66	3.10
Foundry machinery	11.92	7.38	—

TABLE A.3 (*Contd*)

Item	*1974/5*	*1975/6*	*1976/7*
Gas plants	—	0.14	—
Garage equipment	144.91	138.52	89.00
Glass manufacturing plant	16.45	0.91	2.39
Grinding and polishing machinery	12.03	2.18	—
Industrial machinery (n.e.s.)	1.45	23.98	—
Industrial valves	—	86.65	—
Jute mill machinery	38.62	105.67	138.11
Knitting machines	136.92	48.90	18.86
Leather machinery	1.16	1.13	—
Machine tools (all sorts)	712.12	838.77	1692.62
Mechanical handling equipment	76.00	112.19	211.40
Metallurgical plant	—	0.31	—
Mining equipment	—	0.54	10.14
Milk manufacturing plant	—	12.66	14.10
Moulding machinery	33.36	35.25	50.96
Office machinery	521.98	410.94	35.37
Oil mill machinery	122.28	130.10	156.15
Paper and pulp mill machinery	33.69	17.90	205.60
Plastic machinery	6.41	4.48	—
Printing and book binding machinery	10.83	10.97	28.52
Razor blade manufacturing plant	0.42	10.09	—
Reduction gear and parts	37.12	26.67	—
Rice, dal and flour mill machinery	43.04	149.25	115.16
Road rollers	—	0.50	15.83
Rolling mill machinery	3.75	25.23	115.74
Rubber manufacturing plant	2.98	3.56	—
Sewing machines and parts	69.37	71.25	156.20
Shoe making machinery	1.40	4.89	6.09
Sluice valves	35.46	33.12	—
Stone crushing machinery	0.54	6.50	—
Sugar plant and mill machinery	317.12	441.09	226.57
Sugar cane crushers	20.37	9.53	19.40
Tea machinery	86.72	117.93	66.91
Textile mill machinery	1589.72	2124.42	1372.60
Water treatment plant	24.19	21.33	28.43
Weather treatment plant	10.33	35.84	—
Weighing machines	8.73	6.07	19.06
Welding machinery	—	—	14.52
Wire making machinery	7.03	6.71	—
Wood working machinery	3.65	1.16	—
Machinery (n.e.s.)	115.33	128.92	892.31

NOTE
1 lakh = 100,000.
SOURCE
EEPC, *Handbook of Export Statistics 1976–77* (Calcutta, 1978) pp. 112–14.

TABLE A.4 Indian turnkey projects abroad by firm and industry (as of early 1979)

Firm and industry	Nature of project	Destination	Status	Value
I *Traditional sectors*				
Lakshmi Textile Exporters	Textile mills	Malaysia	Completed	n.a.
		Sri Lanka	Completed	n.a.
		Tanzania	Completed	n.a.
Walchandnagar Industries	Sugar mill	Tanzania	Completed	$3 4m
Hindustan Salts Limited	Salt refinery	Tanzania	Completed	$2 m
Simon Carves	Animal feeds	Iraq	Completed	$1 m
Consortium of Indian firms	Cotton spinning	Tanzania	Completed	$12 m
National Small Scale Industries Corporation (Govt of India)	32 Small-scale industries	Tanzania	n.a.	n.a.
Larsen and Toubro	Dairy plant	Yemen	In hand	$1.4 m
II *Electrical*				
Bharat Heavy Electricals Limited (Govt of India)	Power generation (stations, boilers, transmission)	Malaysia	Several projects completed and in hand	$54 m
		Libya	In hand	$122 m
		Libya	Starting	$625 m
		New Zealand	Completed	$12 m
		Saudi Arabia	In hand	$150 m
		Jordan	In hand	n.a.
		Thailand	Starting	$3 m
Tata Exports	Transmission lines, substations, power distribution	Philippines	n.a.	n.a.
		Thailand	n.a.	n.a.
		Algeria	n.a.	$40 m
		Venezuela (with Kamanis)	n.a.	$200 m
		Egypt	n.a.	$14 m

TABLE A.4 (*Contd*)

Firm and industry	*Nature of project*	*Destination*	*Status*	*Value*
Kamani Engineering Corporation	Cables and Transmission towers	Venezuela (with Tata)	n.a.	(see above)
		Iran (with Brown-Boveri)	Implemented	$40 m
		Libya	Implemented	$85 m
		Iran	Implemented	n.a.
Hind Galvanizing & Engng Co.	High voltage transmission	Malaysia	n.a.	$8.2 m.
Voltas	Power generation and distribution	(Several countries details n.a.)	n.a.	n.a.
Electrical Mfg Co.	Transmission lines	Dubai	Completed	n.a.
III *Chemical*				
Balmer Laurie (Govt of India)	Oil refinery structures	Bahrain	n.a.	n.a.
	Oil storage tanks	Burma	n.a.	n.a.
	Fertilizer structures	Sri Lanka	n.a.	n.a.
	Oil terminal	Abu Dhabi	n.a.	n.a.
	Lube blending plant	Abu Dhabi	n.a.	$2 m.
Sarabhai Chemicals	Multipurpose pharmaceutical plant	Cuba	n.a.	n.a.
Engineers India Ltd (EIL) (Govt of India)	Subcontracted engineering, construction	Sri Lanka	n.a.	n.a.

	commissioning (From Kellogg U.S.) for fertilizer plant			
Vijay Tanks and Vessels	Sulphuric acid plant	Kenya	n.a.	n.a.
	Oil storage tanks	Kenya	n.a.	$40 m.
Asian Paints	Paint factory	Yemen and Saudi Arabia	Under negotiation	n.a.
De Smet (Subsidiary of De Smet International)	Solvent extraction plant	Sudan	n.a.	n.a
		Iran	n.a.	n.a.
		Yugoslavia	n.a.	n.a.
		Malaysia	n.a.	n.a.
		Tanzania	n.a.	n.a.
		Philippines	n.a.	n.a.
		Indonesia	n.a.	n.a.
Polychem	Industrial Alcohol	Indonesia	Complete	n.a.
Indian Drugs and Pharamaceuticals Ltd (Govt of India)	Basic drugs and antibiotics	Algeria	Contract awarded	$0.6 m.
		Sri Lanka	n.a.	n.a.
		Afghanistan	n.a.	n.a.
Rubber Reclaim Co.	Rubber factory	Tanzania	n.a.	n.a.
Shalimar Paints	Paint factory	Zanzibar	Complete	n.a.
Engineering Construction Co-op.	Natural gas liquification plant	Doha-Qatar	n.a.	n.a.
IV *Telecommunication and electronic*				
Telecommunication Consultants Ltd	Laying coaxial cables	Libya	n.a.	n.a.
Engineering Projects India Ltd (Govt of India)	Expansion TV, radio station (construction, testing, commissioning sub-contracted from Mitsubishi (Japan))	Iraq	In hand	$20 m.

TABLE A.4 (*Contd*)

Firm and industry	*Nature of project*	*Destination*	*Status*	*Value*
Instrumentation Ltd	Instrumentation and control equipment for power stations	Malaysia	n.a.	$10 m.
Indian Telephone Industries (Govt of India)	Two automatic telephone exchanges	Surinam	n.a.	$2 m.
	Tel. connections	Muscat/Aman	n.a.	$3 m.
	Switching equipment	Jordan	n.a.	$1 m.
	Cable laying	UAE	n.a.	$50 m.
	Railway control equipment	Nigeria	n.a.	n.a.
	UHF equipment	Sri Lanka	n.a.	n.a.
V *Steel, engineering*				
Hindustan Machine Tools (Govt of India)	Machine tool factories	Nigeria Algeria	Joint venture negotiated	
		Kenya Philippines Sri Lanka Iran Iraq Malaysia Indonesia	n.a.	$12 m.
Triveni Structurals Ltd (Govt of India)	Hydraulic structures	Thailand	In hand	$0.3 m.
		Zambia, Tanzania	Complete	n.a.
Sayaji Engng Co.	Rock crushing plant	Afghanistan	Complete	n.a.
		Zambia	Complete	n.a.

Partap Steel Rolling Mills	Steel mill	Mauritius	Complete	n.a.
Consortium of firms	Steel mill	S. Arabia	In hand	
Heavy Engineering Corp. (Govt of India)	Electrolyser pots	Yugoslavia	Complete	n.a.
	Coke ovens	Egypt	Complete	n.a.
	Coke ovens (subcontracted from Russia)	Bulgaria	Complete	n.a.
	Coke ovens continous casting plant	Turkey	n.a.	n.a
	Cranes	Cuba	n.a.	n.a.
Karnataka Implements & Machinery Co.	Farm implements plant	Tanzania	Complete	n.a.
Gammon India (Subsidiary of Gammon UK)	Desalination plant	Dubai	n.a.	$9 m.
Engineering Construction Corp. Ltd.	Natural gas liquification complex	Doha-Qatar	n.a.	n.a.
	Dairy plant	Yemen	n.a.	n.a.
Western India Erectors Ltd	Boilers, turbines,	Kuwait	n.a.	n.a.
	sugar, cement	Iran	n.a.	n.a.
	other plant	Algeria	n.a.	n.a.
	erection (sub-contracting from other firms)	Iraq	n.a.	n.a.
Paharpur Cooling Towers	Air conditioning equipment	Bali	n.a.	n.a.
		Malaysia	n.a.	n.a.
		Kuwait	n.a.	n.a.
Elecon Engineering Co.	Conveyors for sugar terminal	Mauritius	Completed	$3 m

TABLE A.4 (*Contd*)

Firm and industry	*Nature of project*	*Destination*	*Status*	*Value*
Wanson	Industrial boilers	Egypt	n.a	n.a
		Iran	n.a	n.a
		Sri Lanka	n.a	n.a
		Indonesia,	n.a	n.a
		Thailand	n.a	n.a
		Kenya	n.a	n.a
Engineering Projects India Ltd (Govt of India)	Water Treatment plant	Thailand	n.a	n.a
	Part of petrochem. plant	Abu Dhabi	n.a	n.a
	Electrification	S. Arabia	n.a	n.a
	Silos, steel plant	Iraq	n.a.	n.a.
	Coke oven plant	Yugoslavia (with HEC)	n.a.	n.a.
	Fish canning plant	Maldives	n.a.	n.a.

SOURCE
Various newspapers.
N.B.: This list is *incomplete*.

TABLE A.5 India: licensing of patents, designs and know-how abroad (as of early 1979)

Firm	Industry	Destination	Nature of agreement
Indian Drugs & Pharmaceuticals Ltd (Govt of India)	Pharmaceuticals	Afghanistan	Process know-how
		Far East	Licensing
National Research Development Council (Govt.of India)	Leather	US	Patent
	Monochloroacetic acid	UK	Patent
	Spice oleoresins	Malaysia	Know-how
	Construction Materials	Philippines	Process know-how
	Menthol	Nepal	Process know-how
	Solar water still	Saudi Arabia	Process know-how
Amar Dye-Chemicals	Reactive dye mfre.	US	Process know-how
		Brazil	Process know-how
		Sri Lanka (negotiating)	Process know-how
		Algeria (neg.)	Process know-how
		Tanzania (neg.)	Process know-how
Unichem	Anti-inflammatory drug	n.a.	Patent
Godrej Soaps	Soap	Iran	Production licence
Atlas Cycles	Bicycles	Tanzania, Iran, Zambia, Guyana, Sudan, Bangladesh	Production know-how Licence
Tata Engineering & Locomotive Co. Ltd	Commercial vehicles	Indonesia, Egypt, Guyana, UAE, Kenya (?)	Assembly under licence

TABLE A.5 (*Contd*)

Firm	*Industry*	*Destination*	*Nature of agreement*
Bajaj Auto. Ltd	Scooters	Taiwan Indonesia	Assembly under licence
Scooters India (Govt of India)	Scooters	Indonesia Colombia (?)	Assembly under licence
[Mr I. K. Bharati]	Steel	W. Germany	Process design
Bihar Alloy Steels	Continuous casting of alloy steel	W. Germany France	Know-how for adapting Demag (W.G.) equipment
Wanson Pvt. Ltd	Industrial boilers	Canada	Know-how for manufacture of steam boilers
Cooper Engineering	Diesel engines	Bangladesh	Licenced man.
Parle (Exports) Ltd	Soft drinks	Muscat	Franchise
Computronics	Computer software	France	Conversion of programmes (£1 m)

SOURCES
Newspapers, Indian Investment Centre, Department of Science and Technology, Interviews.
N.B.: Data are *incomplete*.

TABLE A.6 Indian engineering consultancy firms active abroad (as of early 1979)

Firm	Industry	Destination	Nature of service
National Industrial Dev. Corporation (NIDC) (Govt of India)	Textile	Algeria	n.a.
	Paper	Guyana	Feasibility study
	Paper	Kenya	Feasibility study
	Fish processing	Maldires	Feasibility study
	Glass manufacture	Sri Lanka	Feasibility study
	Town planning	Tanzania	Planning
	Transport	Italy	n.a.
	n.a.	UK	Project report
	Machine tools, etc.	Iran	various
	Various sectors	Tanzania	various
	Mineral	Afghanistan	Survey work
	Textiles	Iraq	n.a.
	Steel, machine tools	Kuwait	n.a.
	Steel	Libya	n.a.
	Economic planning	Libya	5-Year Dev. Plan
	Starch, sugar, jute	Nepal	Various
	Industrial estate	Guyana	n.a.
Engineers India Ltd (EIL) (Govt of India)	Refineries	Iran	Subcontracted from Snam Progetti (Italy)
	Refineries	Syria	n.a.
	n.a.	Somalia	Subcontracted from Ingeco (Italy)
	Water treatment, refinery, petro-		

TABLE A.6 (*Contd*)

Firm	*Industry*	*Destination*	*Nature of service*
	chemical complex	Iraq	Various
	Various	Somalia, Sri Lanka, Algeria, Bahrain, UAE	
Orient Paper Mills	Paper	Kenya	n.a.
	Paper	Nigeria	n.a.
Birla Textile Mills	Textiles	Sudan	n.a.
Development Consultants	Cement	Iraq	n.a.
	Cement	Libya	Management contract
Tata Consulting Engineers	Power, electrical	Iran	Training and management
	Power, electrical	Kuwait	Installation
	Power, electrical	W. Germany	Design services
	Power, electrical	Malaysia	Construction supervision
	Cotton spinning	Tanzania	Training
	n.a.	Algeria, Sri Lanka, Liberia, Iraq	n.a.
Water & Power Development Consultancy Services (WAPCOS) (Gcvt of India)	Hydro and thermal power	Afghanistan, Burma, Cambodia, Indonesia, Iraq, Laos, Malaysia, Mauritius, Nigeria Nigeria, Philippines, Singapore, Sri Lanka, Tanzania	Design, construction, supervision

Engineering Projects India (EPI) (Govt of India)	Rural electrification	Egypt	Studies
Rail India Technical & Engineering Services (RITES) (Govt of India)	Railways: construction and management	Jordan	Negotiating construction (jointly with two European Firms)
		Nigeria	Managing entire railway system
		Syria, Iran, Zaire, South Korea, South Arabia, Thailand, Libya Philippines, New Zealand	Technical assistance
Fertilizer Corporation of India (Govt of India)	Fertilizers	Burma	Marketing study
		Philippines	Marketing study
Polychem	Industrial alcohol	Kenya, Mauritius, Philippines	Feasibility studies
Rubber Reclaim Co.	Rubber	Sri Lanka	Feasibility study
Central Machine Tools Inst. (Govt of India)	Machine tools	Iran	Study for research institute
INFIN Consult, Pvt. Ltd	Steel mill	Mauritius	Feasibility study
Karnataka Implements	Earthmoving and agricultural implements	Nigeria	n.a.
		Malaysia	n.a.

TABLE A.6 (*Contd*)

Firm	*Industry*	*Destination*	*Nature of service*
Hindustan Machine Tools (HMT) (Govt of India)	Machine tools	Sri Lanka, Philippines, Tanzania, Indonesia, Iran, Kuwait, Malaysia, Algeria	n.a.
Metallurgical & Engineering Consulting Organization (MECON) (Govt of India)	Steel and Aluminium Industry	Nigeria, Bangladesh, UAE, Liberia, Mexico, Hungary	Plant design, Supervision Feasibility studies
Dasturco	Steel	26 Foreign countries	Various
Development Consultants Pvt. Ltd	Sponge iron plants	South Korea, Egypt	n.a.
	Materials handling	Egypt	n.a.
Indian Drugs & Pharmacueticals Ltd. (IDPL) (Govt of India)	Pharmaceuticals	Various Arab countries	Technical assistance for ACDIMA
	Pharmaceuticals	Afghanistan	n.a.
	Pharmaceuticals	Sri Lanka	Feasibility study
Oil and Natural Gas Commission (ONGC) (Govt of India)	Oil drilling and Exploration	Iraq	
		Tanzania	Exploration

SOURCES
Various newspapers, *Economic and Commercial News*, Indian Invt. Centre.
N.B.: This list is *incomplete*.

TABLE A.7 India: Industrial distribution of R & D expenditure in reporting companies 1974[a]

	R & D expen-diture	*Sales*	*Ratio of R & D to sales*	*Share of central govt enterprises*	
				R & D	*Sales*
	(Rs. million)			%	
Chemicals:	203.6	14658.5	1.40	21.5	24.1
(a) Inorganic	12.4	905.9	1.37	—	—
(b) Heavy organic	38.2	3059.1	1.25	86.6	91.1
(c) Synthetic fibres	23.7	2496.7	0.95	—	—
(d) Dyestuffs	17.1	1142.8	1.52	—	—
(e) Synthetic resins and plastics	20.9	1800.3	1.16	—	—
(f) Drugs and pharma-ceuticals	82.5	3623.6	2.28	11.6	14.7
(g) Other	10.5	1630.1	0.64	15.3	13.4
Instruments	7.7	182.8	4.21	57.3	56.3
Electronics and electricals	123.2	10777.3	1.14	47.5	38.1
Machinery	39.1	3322.4	1.18	26.3	38.2
Transport equipment[b]	62.1	3209.0	1.94	75.4	45.2
Office and domestic equipment	33.5	5970.2	0.56	—	—
Metals	30.2	16586.7	0.18	41.7	71.3
Ceramics and glass	1.6	144.5	1.11	0.2	5.3
Cement	7.5	1598.3	0.47	—	—
Paper	1.5	6361.8	0.02	0.7	0.2
	512.0	62819.5	0.82	28.3	35.8

NOTES
[a] Financial year ending 31 March 1974.
[b] Excludes railways which are a departmental undertaking; their R & D expenditure was Rs. 44.6 million.

SOURCE
Desai (1980).

Notes

CHAPTER 1 INTRODUCTION

1. See Lall (1979), Katz (1978b), Katz and Ablin (1978), Cortez (1978), Rhee and Westphal (1978), Lecraw (1977) and Diaz-Alejandro (1977).
2. See Wells (1978), Heenan and Keegan (1979). Smaller countries like the Philippines, Malaysia, Colombia and others are also starting to export their own mini-multinationals, as figures collected by the UNCTC (1978) pp. 231 and 246–7 show.
3. This is discussed in detail by Rosenberg (1976), Freeman (1974) and Katz (1978b).
4. Rosenberg (1976) pp. 146–7. Emphasis added.
5. For a succinct exposition of the elementary principles see Yotopoulos and Nugent (1976) ch. 9.
6. See David (1975) and Nelson and Winter (1977).
7. The 'evolutionary' theory of Nelson and Winter has been developed mainly for process improvements, but it may easily be applied to product adaptations (even its simplification) and improvements. See Harman and Alexander (1977).
8. For a survey of recent literature on Third World manufactured exports see Lall (1980).
9. In some of its empirical formulations, endowments of skill and technology are taken to be a direct function of *per capita* income: thus, a very poor country like India will be taken to export products at the bottom of the skill/technology ladder. See, for instance, Hirsch (1977).
10. For a good reviewer of this literature see Stewart (1977).
11. An excellent case in point is the recent UN Conference on Science and Technology in Vienna.

CHAPTER 2 THE DEFINITION OF 'TECHNOLOGY EXPORTS'

1. I owe this classification to Mariluz Cortez of the World Bank.

CHAPTER 3 SOME BACKGROUND INFORMATION

1. Note that the definition of 'engineering products' is wider than those of the UN (1978). We refer to UN data later in this chapter.
2. The Rupee figures are taken from the EEPC publication and converted to US dollars at current average exchange rates. The dollar values are higher than those given by the UN (1978) and lower than those shown by the World Bank (1977).

3. These data are taken from the World Bank (1977) vol. 1, p. 14. On developing country manufactured exports more generally, and the relatively sluggish performance of India, see Lall (1980).
4. For the top five developed country importers of engineering goods, the percentage in total exports in 1976/7 was 5 per cent for the US, 8 per cent for W. Germany, 3 per cent for the UK, 6 per cent for Holland and 1 per cent for Japan (the USSR had 6 per cent). See Table A.1.
5. However, see Amsden (1979) for a first attempt at statistical analysis of this possibility for a group of developing countries.
6. See Hirsch (1977) for an exposition of this hypothesis in the context of import-substituting industrialization more generally.
7. Information from various issues of *Economic and Commercial News*, Government of India.
8. This has been analyzed in several detailed studies for Argentina by an IDB/ECLA research programme under the direction of Jorge Katz. See, in particular, Katz and Ablin (1977) for the role of 'minor' innovation in comparative advantage.
9. For an analysis of this in Taiwan's case, see Amsden (1977).
10. See World Bank (1977) vol. 2, p. 57.
11. For instance, TELCO trucks, based on old Daimler-Benz designs are a great success in Asian countries; in Malaysia (according to interviews with the firm) they outsell Mercedes trucks.
12. See Helleiner (1973).
13. For evidence on the role of MNCs in developing country exports see Nayyar (1978) and Lall (1980).
14. World Bank (1977) vol. 2, p. 56.
15. Such specialization is, as Rosenberg (1976) argues, one of the major factors behind the growth and technical progress of the US capital goods industry. The benefits of specialization apply particularly to machine tool manufacture where large firms coexist happily with small ones in all industrialized countries. On the problems of developing country producers see Datta-Mitra (1979) and Pack (1978).
16. It is important to note that the UN definition of engineering goods is narrower than the Indian EEPC one; in particular, it excludes metal products which have been the largest and most dynamic element of Indian exports in this field. The figures given in Table 3.4 are not, therefore, comparable with those in Table 3.2 above.
17. According to the World Bank (1977) study of Indian exports (Table 5.5) the share of the top ten exporting firms in engineering exports declined from 28 per cent in 1972/3 to 18 per cent in 1975/6, and of the first fifty firms from 48 to 41 per cent. On the other hand, in Korea, the fastest growing exporter in the world, only a handful of large, integrated trading/manufacturing houses (the *sogo shosha* in Japanese) have spearheaded its fantastic export performance. In 1979, a mere 12 registered *sogo shosha*, with 300 overseas marketing branches, will account for 40 per cent of total exports (see the 'Special Survey' on Korea by *The Financial Times*, London, 2 April 1979, p. 19).
18. For data on Brazil and Mexico see Newfarmer and Mueller (1975) and Lall (1980), for South Korea see Sung Hwan Jo (1976) and on Singapore see Lall (1980). In all these countries, MNCs contributed from two-thirds (Korea being

on the low side) to four-fifths or more (Singapore probably over 95 per cent) of engineering exports. Also see Helleiner and Lavergne (forthcoming) on the role of MNCs in high-technology trade, particularly in the form of intra-firm transactions, of developing countries.

CHAPTER 4 TURNKEY PLANT EXPORTS

1. In this section we include power generation and distribution, telecommunication systems and air-conditioning systems under 'industrial' turnkey projects.
2. The government has played an active role in promoting a consortium approach among Indian enterprises, and public sector manufacturers/contractors have often taken a lead in organizing bids for large projects.
3. See Lall and Streeten (1977).
4. It has been noted for Mexico, for instance, that a number of local firms were acting as 'junior partners' to US firms partly because of their lack of technology but partly also because, even when they possessed the technology a US partner provided them with an established brand name. See Nadal (1977).
5. As noted earlier, slightly older, stable techniques will have a market in the Third World except in industries where the new technique renders *all* older ones uneconomical at all configurations of factor prices and levels of skill availability.
6. The BHEL–Siemens agreement has aroused strong criticism in India (on the grounds that India is 'selling out' to the multinationals) and has not yet been officially approved. It is difficult to understand the reasons for this sort of criticism. Not even the most developed countries hope to achieve technological independence in much of modern industry, and for a poor developing country like India the economic benefits of investing in very advanced R & D (where the results of such R & D may be bought at reasonable prices) are somewhat dubious. On the nature and costs of technological innovation in heavy electricals see Epstein (1972).
7. We are excluding South European countries like Spain and Yugoslavia from the Third World category.
8. A group of 28 Mexican enterprises set up a firm called 'Technimexico' in 1974 to sell technology abroad. According to its brochure (1978) some 3–4 enterprises have done overseas turnkey work in industry, mainly in neighbouring Central American countries. It seems a reasonable assumption that this consortium comprises most of the major private sector technology exporters in Mexico.
9. Rhee and Westphal (1978) p. 12. I have, after discussion with Westphal, excluded 1 'turbine plant' sold to Sweden because this seems to be a mis-specification of an equipment export. I am grateful to Westphal for data on Korea.
10. As the *Financial Times* Survey (2 April 1979) p. 25, notes, in the building of plant for metal, energy and chemical industries, 'the Koreans know that they are at a disadvantage compared with their competitors from Europe and the U.S., who are called in at the beginning of such projects for technical advice and whose specialist companies in such fields as turbine manufacture then take the high value-added portion of contracts'. A similar inference is drawn by Rhee and

Westphal (1978) p. 14. While Indian firms also sometimes subcontract from Western companies, they tend to be passive partners rather than the main contractors.

11. *Ibid.*
12. See Amsden (1977) for an analysis of Taiwanese machinery manufacturers.

CHAPTER 5 DIRECT INVESTMENTS ABROAD

1. On Latin America, see Diaz-Alejandro (1977); on Hong Kong, Wells (1978); on the performance of Third World multinationals as compared to developed country multinationals in Indonesia, see Wells and Warren (1979), and in Thailand see Lecraw (1977); and on the growth of developing country foreign investments in general see Wells (1977) and Heenan and Keegan (1979). For some (incomplete) data on developing country foreign investments see the UN CTC (1978).
2. Ozawa (1979) argues that most Japanese investments abroad have been of the 'mature' variety. This may be so, but many of its current foreign investments in the US and Europe *are* based on innovational leads.
3. To quote from the revised official *'Guidelines Governing Indian Joint Ventures Abroad'*, (Sept. 1978), 'Indian participation should normally be in the form of export of indigenous plant and machinery/equipment required for the joint ventures. However, on merits of each case, participation in one or more of the following forms may also be considered ... (i) export of know-how; (ii) capitalisation of service fees, royalties and other payments; (iii) raising of foreign exchange loans abroad; (iv) grant of loans by Indian participating companies.... Normally, cash remittance will not be allowed for meeting equity contribution but.... deserving cases will be considered on merits and on consideration of the fields of collaboration. For example, cash remittance may be considered in cases of consultancy and other service ventures.'
4. See Wells (1977).
5. Lecraw (1977) describes how Third World investors (mostly Indian) in Thailand achieve better capacity utilization, higher profits and higher reinvestment rates than developed country MNCs.
6. Information provided by officials of TELCO. The original design has been extensively modified by TELCO in both production processes and the vehicle itself, but the vehicle remains much simpler than models made in developed countries.
7. The technological capability of TELCO does not, however, extend to the development of a new engine for a completely different vehicle. The firm has had to purchase a new licence to make the engine for a large (13 ton) vehicle.
8. See Wells (1977) and Lecraw (1977). A future possibility is for Indian direct investors to use 'offshore processing' for Indian materials (e.g. bulk pharmaceuticals) and to have buy-back arrangements. This is presently being discussed for investments in the Sri Lanka free-trade zone.
9. According to the *Financial Times* report cited above, there are 89 new ventures in the 'planning stage' for a value of Rs. 920 m ($52 m). The average value of these ($6 m) is considerably higher than the ones in operation ($3 m), testifying to the entry of investors into larger scale operations.

10. See also Diaz-Alejandro (1977).
11. Much of this information on Hong Kong comes from Wells (1978).
12. See Lall (1979a).
13. It may be noted that the number of overseas trading branches of Korean firms here is much lower than the 200 mentioned by the *Financial Times* Survey of South Korea, 2 April 1979 referred to in Chapter 3 above.

CHAPTER 6 LICENSING

1. A similar point was made in Katz and Ablin (1978) p. 34, about Argentinian technology exports.
2. I am grateful to an official of Hoechst, West Germany, for this information. Also see Parthasarathi (1979) on the sale of patents for an Indian injectible contraceptive to Europe.
3. The *Financial Times* (London, 24 July, 1979) p. 5, reports that Computronics of India has recently signed a contract for around $2.3 million with Centi (France) for conversion work, and is negotiating with other Western firms. British and German firms are particularly anxious to link up with Indian firms, of which about 9 are said to be in the 'take off' stage of programming, to exploit mid-Eastern and far-Eastern markets.
4. Information provided by the Director of Bihar Alloy Steels.
5. The list of firms (158 in number) is contained in the Indian Investment Centre's *India Exports Technology* (New Delhi, 1978). The magazine report states: 'India claims to have boosted its exports of local technology expertise by 45 per cent this year. Along with such multinationals as Dunlop, Johnson & Johnson, Mitsui and Colgate Palmolive, some 60 firms from as far away as Mexico, Australia and Ireland are said to be enquiring about the purchase of Indian knowhow. Areas of interest range from solar water heaters to chemicals, agricultural machinery and engineering.' *8 Days* vol. 1, 17 (London 6 October 1979) p. 25.
6. See 'Wanson goes West', *Business India* (11–24 Dec. 1978) p. 42.
7. We may also note from the table on turnkey projects that the National Small Industries Corporation is setting up several small-scale industries in Africa; this is probably more a form of technical assistance than turnkey work proper.
8. *Business Week* (11 June, 1979) p. 53 reports on major breakthroughs in the licensing of Mexican steel technology in Venezuela, Brazil, Indonesia, Iran, Iraq and Zambia. The construction of these plants is undertaken, however, not by Mexican engineering firms but by firms from the US, Germany and Japan.
9. Katz and Ablin (1978) pp. 5–6, n. 7.
10. *Ibid*., p. 34.
11. *Financial Times* (London, 5 April 1979) p. 5.

CHAPTER 7 CONSULTANCY

1. See the excellent paper by Roberts (1973) on the role of engineering consultants in technology transfer and development. Katz and Ablin (1978) pp. 36–8, provide some additional arguments for the positive effects of consultants on the transfer of technology and the stimulation of complementary exports.

2. For a longer discussion of these factors see Roberts (1973).
3. The Ministry of Railways has recently set up an independent organization, the Indian Railway Construction Company (IRCON) to undertake turnkey construction work in railways. IRCON is now bidding on its own for several contracts abroad. The Projects and Equipment Corporation (PEC) was formed to bid for and export railway equipment.
4. Information kindly provided by executives of MECON and by its balance-sheets.
5. MECON has also been awarded the job of building a complete township to house workers from this steel mill.
6. Information provided by R. Lalkaka of Dasturco.
7. See Diaz-Alejandro (1977).
8. Katz and Ablin (1978) p. 6. The consortium Tecnimexico's brochure also mentions some half-dozen firms with foreign experience of industrial engineering consultancy.

CHAPTER 8 NON-INDUSTRIAL TECHNOLOGY

1. This includes industrial as well as civil projects.
2. As noted before, the metallurgical consultants MECON have also undertaken to build a 4500-house township for the steel plant they are engineering in Nigeria.
3. *Financial Times* (London, 18 April 1979) p. 4.
4. Rhee and Westphal (1978) p. 16. According to the *Financial Times* survey (p. 25) of South Korea (2 April 1979) foreign sales account for half of the total sales of Korean construction industry.
5. See a report by A. S. Frazão 'Brazilian Construction Firms Go Abroad', in the *International Trade Forum* (UN) (April–June 1979).
6. Rhee and Westphal (1978) p. 17. Frazão notes for Brazilian construction firms that their major competitive edge lies in their experience of quickly training labour to work on very large-scale projects.
7. These are presumably all offshoots of the large construction companies, *only 10 of which account for half the Korean industry's turnover*. The remainder is accounted for by 519 companies, all very small (see the *Financial Times* survey just cited). Clearly some small firms also operate abroad under the aegis of the Overseas Construction Corporation.
8. 'Overseas Construction', *Financial Times* (London, 14 March 1978) p. 19.
9. 'Saudi Arabia XV', *Financial Times* (London, 23 April 1979) p. 25. This report also notes that the Korean Overseas Construction Corporation has set up an office in London 'to look for European joint venture partners'.
10. 'Indian Industry', *Financial Times* (London, 14 August 1978) p. 21.
11. 'Saudi Arabia', *Financial Times* (London, 23 April 1979) p. 25. Furthermore, there is 'still no problem of bringing Koreans to the inhospitable climate of the Arab world. Hyundai has a 95 percent reapplication rate from workers who have completed their 13-month contract'. *Financial Times* (2 April 1979) p. 25.
12. *Financial Times* (2 April 1979) p. 25. In a later report (16 October 1979, p. 6) the same paper reports that contracts won by Korea abroad (in all areas) declined by 34 per cent in the first 9 months of 1979 as compared to the corresponding

period in 1978 due to 'growing competition in the Middle East construction market'.

13. The Life Insurance Corporation of India has also set up a branch in East Africa.
14. India has also signed an agreement with Bahrain for the supply of hospitals, vaccine, medicines and for the training of hospital personnel.

CHAPTER 9 NATURE OF EXPORTED TECHNOLOGY

1. See, for instance, Habbakuk (1962), Rosenberg (1972, 1976), David (1975) and Landes (1969).
2. See Nelson and Winter (1974, 1975, 1977) and David (1975), and for earlier work, Kennedy and Thirlwall (1972).
3. See Salter (1966), Kendrick (1973), Griliches (1973), Kennedy and Thirlwall (1972) and Nelson and Winter (1977).
4. See Jewkes, Sawyer and Stillerman (1961), Mansfield (1969), Nelson and Winter (1977), Kamien and Schwartz (1975), Binswanger and Ruttan (1977), Pavitt (1971), Parker (1974) and Freeman (1975).
5. The only attempts so far are by Katz and associates (various), who have applied the concept of 'minor' innovative activity to productivity improvements within *given* technologies in Argentina.
6. See Findlay (1978).
7. Most perceptively by Nelson and Winter (1974, 1975, 1976) and David (1975).
8. See Harman and Alexander (1977).
9. Nelson and Winter (1977) p. 48, emphasis added.
10. Nelson and Winter (1977) p. 47.
11. *Ibid*. p. 51.
12. For an extensive discussion of uncertainty in innovation see Kay (1979).
13. Nelson and Winter (1977).
14. *Ibid*., p. 52.
15. Nelson and Winter (1977) p. 57. Rosenberg gives examples of 'technological imperatives' of this sort: innovations giving rise to bottlenecks in related processes, need for complementary products, and so on.
16. It is noted, for instance, that process innovation is easier than product innovation, since the additional uncertainty of consumer reaction is less for the former. This is of special relevance to developing countries, and we shall return to it later.
17. Harvey (1979) provides an interesting engineering viewpoint on the process of learning during production activity.
18. For a recent survey see Lall (1978).
19. It was pointed out to me during a visit to Bihar Alloy Steels, for instance, that technicians sent out from Europe (to help implement the very modern technology used in this 'mini steel' plant) were much less trained than their Indian counterparts (who were qualified engineers) and were less able to transcend their narrow specialization. The Indians resolved certain technical problems which led to a process improvement which was then licensed back to Europe.
20. This was brought home to me in the Indian truck making firm, TELCO, whose overseas success has been described above. All innovations abroad are

monitored by TELCO research staff, but only those are selected which, after careful evaluation, are likely to appeal to customers sufficiently to outweigh the increased cost. Furthermore, product innovations are relatively few and expensive, and are handled by a separate R & D department, process changes are much more frequent (300–500 changes are made each year requiring some tooling change) and are handled by production engineering departments in each shop floor. As noted earlier, TELCO trucks are made to carry much heavier loads, go on much rougher terrain, go longer without proper servicing, than their counterparts abroad (and than their original design).

21. Information provided kindly by the Chairman of Hindustan Lever, the subsidiary of Unilever.
22. This point is strongly made by Bhagwati and Srinivasan (1975) p. 223.
23. Ashok Leyland, another Indian truck manufacturer, in which British Leyland hold a majority share, had to modify its front-end structures because local steels could not be formed to the original shape.
24. This is being explored by the present author for the Indian automotive industry (commercial vehicles) under the auspices of the UN Centre on Transnational Corporations (the results should be published in 1980). An interesting example can be provided for Ashok Leyland. A local, and much cheaper, supplier for side-frame members could be developed only if these large, rigid steel structures were modified, from a bent shape to a straight one. The latter can be made without special, heavy presses (for the use of which AL would have had to pay heavily), but it can be used in the vehicle only by changing the engine and transmission lay-out. This has been done successfully, and AL is now talking of persuading British Leyland to adopt this modification, which is apparently economical even in British conditions.
25. According to available information, nearly all the Indian technology exporters, particularly of mechanical/engineering technology, have set up R & D departments. Some, like the large public sector units (BHEL, HMT, ITI) have large departments with up to 300–500 engineers; others have smaller ones. They claim to undertake basic design and development work, and to have produced and marketed a large range of completely locally designed products.
26. See Parthasarathy (1979).
27. Note that an enterprise can participate in turnkey exports without reaching this stage if it buys the technological components elsewhere.
28. Note that we are assuming that there is no gap between technical innovation and its commercial application. Such gaps may be important if the innovating institution is separate from the producing enterprise; here, however, we are concentrating on the activities of the latter. We shall return to the 'gap' problem in the next chapter.
29. See Stewart (1977).
30. See Teece (1976).
31. See Lall and Streeten (1977).

CHAPTER 10 DETERMINANTS OF TECHNOLOGICAL DEVELOPMENT

1. However, see Landes (1969) for a challenge to this view.
2. See Lall (1980).

3. For a lengthier discussion see Hirsch (1977). Katz and Ablin (1977) make a similar point about Argentina.
4. There are also related questions of appropriate social structures, attitudes, organizations and the like which may be very important for innovation but which economists are ill-equipped to deal with. For a stimulating discussion, however, see Landes (1969).
5. For a detailed analysis of the case for protecting domestic capital goods manufacture, see Erber's (1978) thesis on Brazil.
6. For a description of the activities of various official bodies in this area and a broader evaluation of Indian technology efforts and plans see NCST (1973), and for a comprehensive review of official measures taken to support local technological activity see the Department of Science and Technology (1978). A useful summary of technological work in India is given by Upadhyaya (1978) and Parthasarathi (1979).
7. For some evidence on India that this is the case see Desai (1980). On more theoretical issues of government intervention in the innovation process see McFetridge (1977).
8. See Department of Science and Technology (1978) pp. 76–7.
9. For a discussion of the importance of uncertainty and risk in the innovation process, and a classification of different types of risk, see Kay (1979).
10. Methods of protecting domestic learning are described below.
11. However, see Bhagwati and Srinivasan (1975) for arguments that technological development may be restricted or biased by a protected, inward-looking environment. Their review of the evidence leads them to conclude, nevertheless, that inward-looking enterprises were not less technically progressive than export-orientated ones, but that the direction of their innovation was different. It would be interesting to see whether these inward-looking innovations discovered technologies which later found markets abroad *because* of the different direction their work had taken.
12. Some of these arguments about R & D by MNCs in developing countries are developed at greater length in Lall (1979b).
13. For a longer discussion of the factors affecting the international allocation of R & D by US MNCs see Lall (1979c).
14. *Ibid*.
15. This phenomenon has become known as the 'truncation' of technology transfer by MNCs in the literature on technological development in Canada. For a thorough discussion of the theory underlying 'truncation', its economic effects and empirical testing (which suggests that truncation does occur, even in as large and developed a host economy as Canada) see McFetridge (1977).
16. According to Upadhyaya (1978) there are over 1.5 million science, technology and engineering graduates in India, which comes third in the world, after the US and USSR, on this score.
17. For a full report on their activities see Department of Science and Technology (1978).
18. According to the NRDC's Annual Report of 1977/78, in the 24 years of its existence 1695 processes have been registered, of which 886 have been licensed and 122 released free of cost. Of these, several have come from manufacturing firms, but the main contributors have been chemical, leather, mechanical engineering, defence, physics and construction material laboratories. It is

estimated that by 1977/78 the value of production from processes licensed by NRDC came to Rs. 645 million (about $80 m).

19. There has also been severe criticism within the country about the usefulness of the expenditure on 'basic' research divorced from manufacturing enterprises, but we cannot go into this problem here. See Upadhyaya (1978).
20. The development of technology for small-scale units has been an explicit aim of manufacturing R & D in the national laboratories.
21. The *timing* of the attack on foreign markets may be explained by a shift to a more outward-looking policy and recession in home markets, but this is quite compatible with the argument that the building up of the capabilities exported required protection.
22. See Parthasarathi (1979).
23. To take the Hindustan Lever example again, a major product innovation, which lightens the colour of the skin, will be produced and marketed abroad, via Unilever headquarters, by the respective subsidiaries. Its machine-building know-how is similarly being transferred to Brazil within the MNC framework, and several of its process innovations are used by Unilever affiliates in South East Asia. These are circumstances where it is profitable for the parent company to use an innovation made in India in the existing structure of operations elsewhere. In others, however, where the innovation is based on specifically Indian materials, or the technology deeply embedded in Indian personnel, it may lead to increased exports (of products, not of technology) from India. An outstanding example is H. Lever's export of sal fat (a cocoa butter substitute), which was developed in India, and is now an important foreign exchange earner. This does not, however, weaken the main argument that given innovations will tend to lead to greater technology exports if they take place within local firms rather than multinationals.
24. See a detailed and perceptive analysis of the Brazilian capital goods sector by Erber (1978), who demonstrates how the lack of protection of basic design capability in local manufacturers led to a decline in their technological ability and to a loss of market shares to MNCs. A similar case is made about Mexico by the World Bank (1979), though not with specific reference to local or foreign enterprises.
25. Evidence on this is provided for the Andean countries by Mytelka (1978).
26. On the crucial significance of design and engineering capability in the heavy electrical industry see Epstein (1972), who argues the bulk of technological costs are involved in 'tooling and manufacturing engineering' rather than advanced research.
27. See Newfarmer (1979).
28. In some industries this would involve learning different processes (e.g. in chemicals or assembly line techniques); in others it may involve learning to operate processes of a certain *scale* (e.g. in power generation technical progress takes the form of increasing the capabilities of machines without changing the basic engineering processes).
29. See Hirsch (1977). For a survey of recent literature see Lall (1980) and for earlier literature see Stern (1975).
30. For this interpretation of the Argentine experience see Katz and Ablin (1977).
31. See Amsden (1979). There may be exceptions, of course, such as Hindustan Machine Tools exporting numerically controlled machines to the US and UK.

References

Amsden, A. (1977), 'The Division of Labour is Limited by the Type of Market: The Case of the Taiwanese Machine Tool Industry', *World Development*, pp. 217–34.

—— (1979), 'The Industry Characteristics of Intra Third World Trade in Manufactures', Columbia University mimeo, *Economic Development and Cultural Change* (forthcoming).

Bhagwati, J. N. and Srinivasan, T. N. (1975), *Foreign Trade Regimes and Economic Development: India* (New York: NBER).

Binswanger, H. P. (1974), 'A Microeconomic Approach to Induced Innovation', *Economic Journal*, pp. 940–58.

Binswanger, H. P. and Ruttan, V. W. (1977), *Induced Innovation: Technology, Institutions and Development* (Baltimore: Johns Hopkins University Press).

Cooper, C. (1974), 'Science Policy and Technological Change in Underdeveloped Economies', *World Development*, pp. 55–64.

Cortez, M. (1978), 'Argentina: Technical Development and Technology Exports to Other LDCs' (World Bank mimeo).

Datta-Mitra, J. (1979), 'The Capital Goods Sector in LDCs: Is There a Case for State Intervention?' (World Bank mimeo).

David, P. A. (1975), *Technical Change, Innovation and Economic Growth* (London: Cambridge University Press).

Department of Science and Technology (1978), *Report 1977–78* (New Delhi: Government of India).

Desai, A. V. (1980), 'The Origin and Direction of Industrial Research and Development in India', *Research Policy* (forthcoming).

Diaz-Alejandro, C. F. (1977), 'Foreign Direct Investment by Latin Americans', in T. Agmon and C. P. Kindleberger (eds), *Multinationals from Small Countries* (Cambridge, Mass.: MIT Press).

Engineering Export Promotion Council (1978), *Handbook of Export Statistics 1976–77* (Calcutta: EEPC).

Epstein, B. (1972), 'Power Plant and Free Trade', in D. Burn and B. Epstein, *Realities of Free Trade: Two Industry Studies* (London: Allen & Unwin).

Erber, F. (1978), 'Technological Development and State Intervention: A Study of the Brazilian Capital Goods Industry' (University of Sussex Ph.D. Dissertation).

Findlay, R. (1978), 'Relative Backwardness, Direct Foreign Investment and the Transfer of Technology', *Quarterly Journal of Economics*, pp. 1–16.

Frankena, M. (1973–4), 'The Industrial and Trade Control Regime and Product Design in India', *Economic Development and Cultural Change*, pp. 249–64.

Freeman, C. (1975), *The Economics of Industrial Innovation* (Harmondsworth: Penguin).

Griliches, Z. (1973), 'Research Expenditures and Growth Accounting', in B. R. Williams (ed.), *Science and Technology in Economic Growth* (London: Macmillan; New York: Halsted Press).

Grubel, H. G. and Lloyd, P. J. (1975), *Intra-Industry Trade* (London: Macmillan).

Habakkuk, H. J. (1962), *American and British Technology in the 19th Century* (London: Cambridge University Press).

Harman, A. and Alexander, A. J. (1977), 'Technological Innovation by Firms: Enhancement of Product Quality' (Santa Monica: Rand Corporation) R-2237-NSF.

Harvey, R. A. (1979), 'Learning in Production', *The Statistician* (Mar.) pp. 39–57.

Heenan, D. A. and Keegan, W. J. (1979), 'The Rise of Third World Multinationals', *Harvard Business Review*, (Jan. – Feb.) pp. 101–9.

Helleiner, G. K. (1973), 'Manufactured Exports from Less-Developed Countries and Multinational Firms', *Economic Journal*, pp. 21–47.

Helleiner, G. K. and Lavergne, R. (forthcoming), 'Intra-Firm Trade and Industrial Exports to the United States', *Oxford Bulletin of Economics and Statistics*, Special Issue on Multinationals, ed. by S. Lall.

Hirsch, S. (1977), *Rich Man's, Poor Man's and Every Man's Goods* (Tübingen: J. C. B. Mohr).

Jewkes, J., Sawyer, D. and Stillerman, R. (1961) *The Sources of Invention* (London: Macmillan; New York: Norton).

Kamien, M. and Schwartz, N. (1975), 'Market Structure and Innovation: A Survey', *Journal of Economic Literature*, pp. 1–37.

Katz, J. (1976), *Importacion de Technologia, Aprendizaje e Industrializacion Dependiente* (Mexico: Fondo de Cultura Economica).

—— (1978a), 'Creacion de Technologia en el Sector Manufacturero Argentina', *El Trimestre Economico*, pp. 167–90.

—— (1978b), 'Technological Change, Economic Development and Intra, and Extra Regional Relations in Latin America', Buenos Aires: IDB/

ECLA Research Programme in Science and Technology, Working Paper no. 30.

Katz, J. and Ablin, E. (1977), 'Technologia y Exportaciones Industriales', *Desarrollo Economico*, pp. 89–132.

—— (1978), 'From Infant Industry to Technology Exports: The Argentine Experience, in the International Sale of Industrial Plants and Engineering Works', Buenos Aires: IDB/ECLA Research Programme in Science and Technology, Working Paper no. 14.

Kay, N. M. (1979), *The Innovating Firm* (London: Macmillan).

Kendrick, J. (1973), *Post-war Productivity Trends in the United States, 1948–1969* (Princeton University Press, for NBER).

Kennedy, C. and Thirlwall, A. P. (1972), 'Surveys in Applied Economics: Technical Progress, A Survey', *Economic Journal*, pp. 11–72.

Lall, S. (1978), 'Transnationals, Domestic Enterprises and Industrial Structure in Host LDCs: A Survey', *Oxford Economic Papers*, pp. 217–48.

—— (1979), 'Developing Countries as Exporters of Technology: A Preliminary Assessment' in H. Giersch (ed.), *International Resource Allocation and Economic Development* (Tübingen: J. C. B. Mohr).

—— (1979a), 'Multinationals and Market Structure in an Open Developing Economy: The Case of Malaysia', *Weltwirtschaftliches Archiv*, pp. 325–48.

—— (1979b), 'Transnationals and the Third World: The R & D Factor', *Third World Quarterly* (July) pp. 112–18.

—— (1979c), 'The International Allocation of Research Activity by US Multinationals', *Oxford Bulletin of Economics & Statistics* (Nov.).

—— (1980), 'A Survey of Recent Trends in Manufactured Exports by Newly Industrializing Countries', in S. Lall, *Developing Countries in the International Economy* (London: Macmillan).

Lall, S. and Streeten, P. P. (1977), *Foreign Investment, Transnationals and Developing Countries* (London: Macmillan).

Landes, D. S. (1969), *The Unbound Prometheus* (London: Cambridge University Press).

Lecraw, D. J. (1977), 'Direct Investment by Firms from Less Developed Countries', *Oxford Economic Papers*, pp. 442–57.

Mansfield, E. (1969), *The Economics of Technological Change* (London: Macmillan).

McFetridge, D. F. (1977), *Government Support of Scientific Research and Development: an Economic Analysis* (University of Toronto Press).

Mytelka, L. K. (1978), 'Licensing and Technology Dependency in the Andean Group', *World Development*, pp. 447–59.

Nadal, A. (1977), 'Multinational Corporations and the Transfer of Technology: The Case of Mexico', in D. Germidis (ed.), *Transfer of Technology by Multinational Corporation*, vol. 1 (Paris: OECD).

National Committee on Science and Technology (1973), *An Approach to the Science and Technology plan* (New Delhi: Government of India). Abridged form published in *Minerva* (1973) pp. 537–70.

Nayyar, D. (1978), 'Transnational Corporations and Manufactured Exports from Poor Countries', *Economic Journal*, pp. 59–84.

Nelson, R. R. and Winter, S. G. (1974), 'Neoclassical versus Evolutionary Theories of Economic Growth: Critique and Prospectus', *Economic Journal*, pp. 886–905.

——(1975), 'Factor Price Changes and Factor Substitution in an Evolutionary Model', *Bell Journal of Economics*, pp. 466–86.

Nelson, R. R. and Winter, S. G. (1977), 'In Search of Useful Theory of Innovation', *Research Policy*, pp. 36–76.

Nelson, R. R. Winter, S. G. and Schuette, H. L. (1976), 'Technical Change in an Evolutionary Model', *Quarterly Journal of Economics*, pp. 90–118.

Newfarmer, R. S. (1979), 'TNC Takeovers in Brazil: The Uneven Distribution of Benefits in the Markets for Firms', *World Development*, pp. 25–44.

Newfarmer, R. S. and Mueller, W. F. (1975), *Multinational Corporations in Brazil and Mexico: Structural Sources of Economic and Non-Economic Power*, report to the US Senate (Washington, DC: Government Printing Office).

Ozawa, T. (1979), 'International Investment and Industrial Structure: New Theoretical Implications from the Japanese Experience', *Oxford Economic Papers*, pp. 72–92.

Pack, H. (1978), 'The Capital Goods Sector in LDCs: A Survey' (World Bank mimeo).

Parker, J. E. S. (1974), *The Economics of Innovation* (London: Longman).

Parthasarathi, A. (1979), 'India's Efforts to Build on Autonomous Capacity in Science and Technology for Development', *Development Dialogue*, pp. 46–59.

Pavitt, K. (1971), *Conditions of Success in Technological Innovation* (Paris: OECD).

Rhee, Y. and Westphal, L. E. (1978), 'A Note on Exports of Technology from the Republics of China and Korea' (World Bank mimeo).

Roberts, J. (1973), 'Engineering Consultancy, Industrialisation and Development', in C. Cooper (ed.), *Science, Technology and Development* (London: Frank Cass).

Rosenberg, N. (1972), *Technology and American Economic Growth* (New York: Harper).

—— (1976), *Perspectives on Technology* (London: Cambridge University Press).

Salter, W. E. H. (1966), *Productivity and Technical Change* (London: Cambridge University Press).

Stern, R. M. (1975), 'Testing Trade Theories', in P. N. Kenen (ed.), *International Trade and Finance: Frontiers of Research* (London: Cambridge University Press).

Stewart, F. (1977), *Technology and Underdevelopment* (London: Macmillan).

Sung-Hwan Jo (1976), 'The Impact of Multinational Firms on Employment and Income: The Case Study of South Korea' (Geneva: ILO, World Employment Programme) WP. 12.

Teece, D. J. (1976), *The Multinational Corporation and the Resource Cost of International Technology Transfer* (Cambridge, Mass.: Ballinger).

UN (1978), *Bulletin of Statistics on World Trade in Engineering Products, 1976* (New York).

UNCTC (1978), *Transnational Corporations in World Development: A Re-examination* (New York).

Upadhyaya, K. K. (1978), 'Design, Development and Research Activity in India', *Lok Udyog* (Mar.) pp. 37–41.

Wells, L. T. (1977), 'The Internationalization of Firms from Developing Countries', in T. Agmon and C. P. Kindleberger (ed.), *Multinationals from Small Countries* (Cambridge, Mass.: MIT Press).

—— (1978), 'Foreign Investment from the Third World: The Experience of Chinese Firms from Hong Kong', *Columbia Journal of World Business* (Spring) pp. 39–49.

Wells, L. T. and Warren, V. (1979), 'Developing Country Investors in Indonesia', *Bulletin of Indonesian Economic Studies* (Mar.) pp. 69–84.

World Bank (1977), *India: Export Performance, Problems, Policies and Prospects* no. 1352-IN (Washington, DC).

—— (1979), *Mexico: Manufacturing Sector: Situation, Prospects and Policies* (Washington, DC).

Yotopoulous, P. and Nugent, G. (1976), *Economics of Development: Empirical Investigations* (New York: Harper & Row).

Author Index

Subject Index